30 Advances in Biochemical Engineering/Biotechnology

Managing Editor: A. Fiechter

Bioprocess Parameter Control

With Contributions by
P. Agrawal, A. Fiechter, C. Kleinstreuer,
H. C. Lim, T. Poweigha, S. Shimizu,
H. Ursprung, T. Yamané

With 88 Figures and 28 Tables

Springer-Verlag
Berlin Heidelberg GmbH 1984

ISBN 978-3-662-15283-6 ISBN 978-3-540-39004-6 (eBook)
DOI 10.1007/978-3-540-39004-6

Originally published by Springer-Verlag Berlin Heidelberg New York in 1984.
Softcover reprint of the hardcover 1st edition 1984

Library of Congress Catalog Card Number 72-152360

2152/3020-543210

Table of Contents

Editorial

Prof. Dr. Heinrich Ursprung is president of the Swiss Federal Institute of Technology in Zurich (ETHZ). In this position, he is responsible for maintaining and developing the standards of this educational and research institution.

Along with the departments of engineering, chemistry, pharmacy, forestry, agriculture, mathematics and physics, a faculty of sciences is also a part of the ETHZ. The foresighted view of President Ursprung made it possible to develop within this latter department, a special curriculum and thus an additional degree in biotechnology. Accordingly, a new institute for biotechnology was implemented at the satellite campus Hönggerberg, as a neighbor to the already existing Institute of Cell Biology and Institute of Molecular Biology.

The rearrangement of the existing faculties of an academic instituion is generally not easy to perform, especially in biology where scientific progress is quite rapid and modifications can be effected only with some delay. Also, situations at most universities are usually not financially favorable to the promotion of quickly developing areas.

As a former cell biologist, Prof. Ursprung very early recognized when molecular sciences had grown to a level for potential application in industry and the environment. He sees that engineering elements have to join the biological components in order to furnish a sound basis for the applied area of biotechnology. Using his vast experience as an academic leader, he has expressed his views in numerous lectures, conversations, and article publications, thus becoming one of the most renowned leaders of science policy.

A. Fiechter

Biotechnology: The New Chance for Industry

Heinrich Ursprung
President, Swiss Federal Institute of Technology (ETH) Zürich, Switzerland

Generous support has been pouring into research on basic biology for decades. It is not surprising that the question of its pay-off is asked with increasing insistence. In other fields it has been customary to relate important technological breakthroughs to cognate discoveries in basic science: this or that discovery, it is said, „*was valorized*". When will this happen in the Life Sciences? When and how will the epoch-making discoveries of molecular, or "new" biology be valorized?

When we call interferons as witnesses, we are just scratching the surface of a valorizing potential of unprecedented dimensions. One must realize what all has become feasible by the discoveries of New Biology. (This term refers not only to the double helix, gene-splicing, and monoclonal antibodies. It refers to the very introduction of quantitative, molecular, and physical thinking into the Life Sciences; to our vast knowledge of enzymology and more generally speaking, protein chemistry; to the spectacular advances in immunology). *The key to valorizing these discoveries and insights is biotechnology*. Its basic strategy is to scale up biological processes to industrial dimensions, by adapting methods borrowed from process and control engineering to biological processes. In this sense, biotechnology is not a natural science, but an engineering science. Its finality lies not primarily in explaining and understanding phenomena, but in producing things. A biotechnologist working on the synthesis of insulin from human genes that were incorporated into bacteria is not primarily interested in DNA surgery, but in industrial scale production of a better insulin without having to rely on pancreas glands. He may want to scale up the manufacturing process to the dimension of a 10'000 gallon fermenter. It should be noted that in the process as a whole, the costs at the DNA surgery end amount to perhaps 5%. The subsequent steps of scaling-up and purification devour 95% of the financial effort. From the economical viewpoint it is therefore imperative to keep improving the methodology of the latter.

It is true that the cases developed in recent years are relatively few: insulin, interferons, vaccines, growth hormones, to mention four. But our knowledge of metabolism is vast, and there is no a priori reason to believe that all enzyme (and therefore, gene) mediated reactions could not be scaled up by fermentation technology, enzyme immobilization, and related techniques. Using enzymes derived from thermostable organisms, heat denaturation problems encountered e.g. in the application for waste-water treatment, can be overcome. By protein engineering, including the chemical synthesis of novel genes, custom tailored enzymes with novel properties can be constructed. There appears to be no theoretical limit. Protein engineering may also lead to new approaches in extreme miniaturization, in a yet ill-defined area of so-called biochips. In fact, chemistry-derived biology and physics-derived electronics may converge into bioelectronics, opening new markets for electronic devices carrying biological specifity, as bioelectrodes or biosensors. The potential application in biomedical technology is vast.

It is self-evident that biotechnology is important for the fine-chemical industry. It is also important for the pharmaceutical industry, not only in view of novel pathways of synthesis, but of novel strategies for drug administration. Liposomes with built-in antibodies against surfaces of target cells can carry their drugs to the right cells when introduced into the bloodstream of a patient. The obvious advantage of this strategy lies in the lower doses of drugs needed, and the concomitant lowering of unwanted side effects.

Biotechnology will have an increasing impact on agrochemistry, too. Plant cell culture techniques have been improved, and mass production of genetically defined plantlets is becoming feasible; it will interest the seed industry. If crop plants resistant to drought or disease can be produced in large quantities cheaply, their impact on agriculture can be substantial. Other developments in crop science concern nitrogen fixation by grasses including wheat and corn. Hopes are high that plant physiologists and geneticists will succeed in enabling the roots of such crop plants to form symbioses with soil bacteria, making them autonomous for nitrogen fixation, imitating legume plants.

Biotechnological thinking has also entered the field of fuels. Will fuels be produced by converting biomass on a large scale? Recent improvements in the degradation rate of lignin look encouraging, as does the conversion of organic waste into precious oxychemicals. Moreover, biotechnology is useful for a more efficient exploitation of conventional fuel sources. Microorganisms that release surfactants are being used for altering the adhesion between sand grains and oil droplets. This example, too, shows that biotechnology is, because of its finality, an engineering science: the objective in this case is not to understand the physiology of surfactant release, but to exploit the biological phenomenon for large scale technical application.

In parallel to the development of techniques for scaling up, an instrumentation industry is growing. Companies with expertise in chemical plant layout or the construction of process equipment are diversifying their efforts into bioreactor design; among the difficulties encountered are problems of sterilization. But the opportunities are large, and they include not only reactor vessels, but entire control systems with intricate sensors measuring those parameters that are crucial for optimizing the process for high yields.

New scientific approaches of the dimension of biotechnology inevitably raise questions of policy. While professors of chemistry or electrical engineering have a long record of fruitful interactions with industry, for many biology professors the opportunity is only recent. Old questions arise anew. Should biotechnology professors serve as consultants to industry? Or does this constitute treason against academia? Should they found companies of their own? Who provides the capital? Should industry, or should the University, create an "interface" between the two? On or off campus, as Science Parks or Research Parks? If a state-operated University enters the field: is it distorting competition? Should it? What becomes of the potential income from such activities? Does it go into the pocket of the professor, his University, the State, or industry? What in biotechnology may be patented? Organism? Genomes?

Whatever the questions, the answers should come quickly. Delays could be very costly to the economy. In several countries, particularly Japan and the United States, hundreds of new companies have been created in recent years in the area of biotechnology. Many of them enjoy the expertise of scientists who left their Universities. In other countries this approach has not been chosen to any extent thus far. There, the problems of transferring biotechnological know-how from Universities to practice remain to be solved. Of course, the normal way of this transfer is to train students and release them from the University. This is a slow process, however, and it functions only if Universities are sufficiently quick in establishing the necessary curricula. Those that have done this may expedite the transfer by offering training courses for staff now employed in industry. This requires decisions on the part of industry to move in the direction of biotechnology. My observations indicate that this, too, is less the case in Europe than in Japan and the United States.

It is nevertheless encouraging to note that in Europe, both science promotion agencies and University authorities are increasingly aware of the vast potential of biotechnology. Some of them have begun to support it substantially. Supporting biotechnology in itself means making an early pay-off more likely. Support biotechnology!

Physical and Chemical Parameters of Microbial Growth

Armin Fiechter
Department of Biotechnology, Swiss Federal Institute of Technology,
ETH-Hönggerberg, CH-8093 Zürich

The observable behavior of a living cell in submerged culture results from the nature of underlying genetic information and its expression. Any expression of the genes is a matter of enzyme activity, which in turn is influenced by numerous factors of metabolic control mechanisms. Besides intracellularly located parameters, environmental factors also play an important role in the actual performance of microbes during growth.

The present review deals with those parameters that act on growth due to mechanisms working outside of the cell such as medium components and the methods which make them available for uptake. The numerous environmental effects can best be divided in the two categories of physical and chemical parameters.

The first category includes temperature, pressure, and the physical nature of the reaction mixture, including the problems of mixing and aeration. Emphasis is therefore given to the configuration of physical containments used in a bioprocess. The ultimate goal of any containment is the supply of nutrients to and the removal of metabolic products from the cell and the maintenance of a uniform distribution of the liquid, gaseous, and solid phases involved in the reaction mixture. Some data are also given on new principles for agitation and aeration. The advantages and drawbacks of the classical stirred tank and the air lift reactor are mentioned and these reactors are compared to three different loop forms used in research and partly in production plants. The tendency to further develop loop forms is becoming apparent, as these allow much better control of the flow pattern, irrespective of the viscosity, the uniform distribution of components, and an optimal supply of nutrients.

The second category of extracellular parameters includes the effects of medium containing substrate, nutrients, growth factors and trace elements. The proper selection of these components and their quantities is of great importance due to their potential effects on the metabolic performance of the cell. Continuous culture methods are shown to be of high efficiency for the fast identification of these effects, and a systematic concept for medium design is developed. Such a concept can replace the troublesome and arduous medium 'optimization' work based on trial and error. It allows for an exact evaluation of regulatory patterns at various growth rates and is therefore most suitable for metabolic studies and investigations of product formation.

The importance of medium designing and the development of appropriate hardware for growth is illustrated by some examples from studies using yeasts and bacteria. It is concluded that improved knowledge of the influence of physical and chemical parameters is prerequisite in any consistent work on metabolism and process development.

1 Introduction

1.1 Phenomena of Microbial Growth

Microbial growth behavior and metabolic control effects are the result of interacting intra- and extracellularly located effectors. Thus, the physical and chemical parameters involved, as well as the genetically encoded characteristics, influence the ultimate cell behavior.

This article deals with an aspect that has been neglected too much in the past — the proper interpretation of the phenomena that occur during growth. This interpretation is, in fact, a problem of the correct identification of all of the effectors involved in a bioprocess. The observable results in such a bioprocess in function of time are represented by the overall effects of growth, e.g., product formation, metabolic activity and its control, uptake of oxygen, release of carbon dioxide, as well as activities of other gases. Also, the specific uptake rates of substrates (q_s) and the release of metabolites like pyruvate, aldehyde, ethanol, acetic acid, etc., or secondary metabolites may depend heavily upon agitation efficiency, substrate (e.g., glucose or oxygen) supply or the availability of nutrients. Especially repressive effects of entire metabolic sequences (e.g., TCA-cycle or electron transfer chain) can be found as a result of these external parameters, as well as as a result of intracellularly localized control.

It is difficult to develop methods that allow for systematic and efficient identification of all underlying effectors and for an accurate sorting out of their adhearing effects from overlapping and superpositioning phenomena. As the number of relevant effectors can be very high, it is necessary to develop an efficient, systematic procedure that does not depend only upon empirical approaches in the identification of single parameters. The relevant aspects are outlined in Sect. 2 where some of the more recent devices for agitation/aeration are discussed and in Sect. 3 on the chemical parameters where some recent data of nutritional studies are given.

Principally, growth can be measured as the increase of cell mass or cell number. Consequently, yield (Y), specific growth rate (μ) and saturation constant (K_s) are the major characteristics of a growth process. In addition, secretion of metabolites, gas exchange and the formation of extracellular and/or intracellular end products complete the full picture of what happens after nutrients and a living cell (able to proliferate) have been combined. The quantitative estimation of the amounts of cells and products allows for the calculation of mass balances and, in function of time, of appropriate turnover rates, which leads finally to a sound basis for the identification of all growth and regulatory phenomena. Thus, the type of growth (exponential, 'linear'), growth relation of product formation, inhibition/repression effects, nutritional or inductional effects can be ascribed to their initiating effectors.

1.2 Effectors of Microbial Growth

Two main categories of effectors can affect a growing system. They are of *intracellular* and *extracellular* origin. The *intracellularly* located potential, which is responsible for cell performance, consists of the genome and the apparatus for replication, transcription, and translation. Gene expression, however, is only partially performed by the genome itself. As all of its steps are also catalyzed by enzymatic actions, interference of extracellular effects during growth may become as significant as the genetic information as such. Under the conditions of application, the genome is taken as a constant basis that does not change as long as the strain used is not replaced by any other type of cell. Therefore, genetic aspects will not be treated in the following sections. Also, the possible influence of spontaneous mutations or the presence of prophages will not be covered, despite their potential importance in investigations of gene expression in mass culture.

The *extracellular* parameters can be divided into two subcategories:

Physical

Prerequisite for the adequate provision of nutrients to the cell is the removal of liquid and gaseous products and the maintenance of necessary physical conditions such as temperature, viscosity and pressure.

Ideal mixing of the solid (cells), liquid (substrate), and gaseous (e.g., oxygen) phases is therefore of paramount importance, irrespective of the physical nature (viscosity, surface or boundary tensions) or the chemical nature of the reaction mixture. Agitation and aeration therefore have to provide adequate mass and heat transfer and keep the respective concentration and energy profiles minimal. Ideal mixing must be achieved as much as possible. Besides these basic aspects, the design of a suitable physical containment must also meet the practical requirements for sterilization, prevention of wall growth, and segregation of the reaction components (foam control), irrespective of the physical nature of the reaction mixture.

Chemical

Growth behavior is predominantly under the control of extracellular nutrients like carbon and energy sources, nitrogen, phosphorus, potassium, manganese, etc. Also, trace elements like heavy metals and growth factors (vitamins, amino acids, etc.) may dictate the proliferation of cells or the formation of products, as well. Besides these four categories of carbon and energy sources, nutrients, growth factors and trace elements, appropriate H^+-ion concentrations and redox properties must be provided. These components are normally dissolved in water and combined, making up a medium that allows exponential or linear reaction kinetics. Depending upon their chemical properties and concentrations, these medium components can exert rather drastic regulatory (inhibition, repression/derepression) or other effects (acidification, precipitation, etc.).

The influence of the chemical components in a growth experiment is quite often underestimated because incompleteness in supplementation of a nutrient does not necessarily stop the related metabolic activity completely. However, the phenomena actually observed may become very atypical, such as high values for the saturation constant k_s of the Monod equation, poor growth rates and poor yields. Unusually high k_s values can indicate both categories of effects, e.g., inhibition and limitation (unavailability of nutrients or nonaccessibility).

Correct medium design is therefore most important for experimental and practical purposes, especially when metabolic effects must be understood. Due to the large number of media components necessary for full growth, design of an appropriate composition, including demonstration of built-in nutrient limitations, requires its own methodology. Thus, Sect. 3 deals not only with the composition of media, but also with efficient ways for their construction.

1.3 Identification and Interpretation of the Nature of Effectors

Because of the large number of parameters that may control the growth of a living cell, identification of the relevant factors is the primary goal for the elaboration of a sound experimental background. Therefore, efficient concepts for determining major influences and their interdependencies must be available.

The basic question for process improvement is how to identify the nature of the parameters that influence the process. The use of a purely empirical approach in the identification and evaluation of potential effectors, based upon trial and error, is no longer acceptable even for apparently simple systems, especially since this approach can be very time-consuming and economically unfeasible. Most important, the chances are very high that even after prolonged efforts — after an extensive amount of time spent in detailed research — the true limitations can still remain unknown. In fact, it is even possible that an artificial system might emerge. Not only is this technique not optimal, but it can lead to inaccuracies.

The first step in determining an efficient identification concept is the categorization of observable phenomena into the three classes of physical, chemical, and biological effectors. Investigators often tend to ascribe things that cannot be easily interpreted to genetically encoded biocontrol and thus overlook the possibility of extracellular factors. This problem of identification cannot always be solved in a systematic way. However, some possibilities are now available through the use of continuous culture methods and their modifications.

An example of the highly informative results of continuous culture methods is given in Fig. 1a. The diagram simulations show x (biomass), s (substrate) and r_x

(productivity) in function of D (dilution rate) based on Monod kinetics. Such a set of data represents a basis for kinetic calculations (q_s, q_p, q_{O_2}, RQ), as well as for mass balances. In Fig. 1a, the curves of x and s are given from a simulation experiment with the saturation constant K_s as a parameter. Calculations are made on the basis of the Monod equation: a rather drastic change in the shape of the curves for x and s becomes visible with increasing values of K_s, yielding atypical forms for growth inhibition, unknown nutrient limitations or other effects. In interpreting this figure,

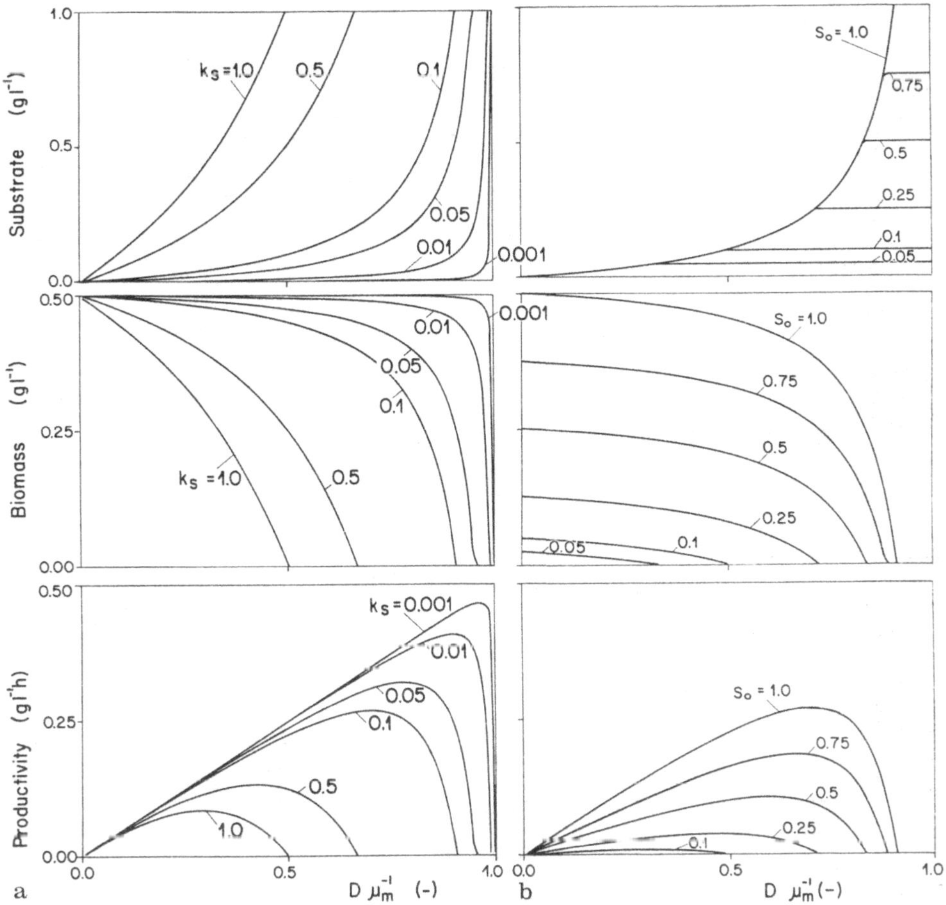

Fig. 1a. Simulation of plots for substrate (s), biomass (x) and productivity (r_x) values in function of dilution rate (D) based on Monod kinetics. *Parameter is the saturation constants* K_s. The characteristics of the plots may be useful for interpretation of experimental data. High values of K_s are always suspect, indicating the presence of inhibitor, poorly degradable substrate(s) or some hidden nutrient limitation

$$\mu = \mu_{max} \frac{s}{k_{s+s}}$$

Relative dilution rate $= D\mu_{max}^{-1} = 1.0\ h^{-1}$, $s_0 = 1.0\ g\ l^{-1}$, $Y = 0.5$; **b.** Computer simulation of the same chemostat experiment based on Monod kinetics as in Fig. 1a. *Parameter is the substrate concentration* s_0.

Relative dilution rate $= D\mu_{max}^{-1} \cdot K_s = 0.1\ g\ l^{-1}$; $Y = 0.5$

it is clear that high values for K_s will always be suspicious, indicating one of the following effects:

- Substrate is poorly degradable
- Presence of an inhibitor (substrate inhibition, production inhibition, . . .)
- Not all cells can divide because of biological defects
- Incomplete distribution (poor agitation, wall growth, formation of foamy overlay, . . .)
- Unknown limitations

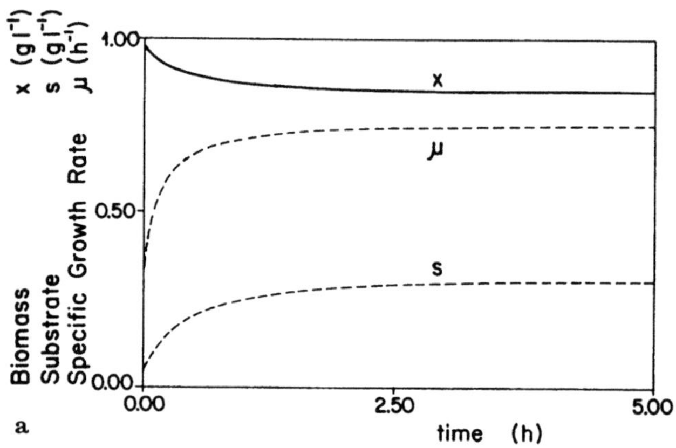

Fig. 2a. Simulation of the transient states (in function of time) in a D-shift (dilution rate) experiment. D = 0.25 → 0.75. x, µ, s in function of time. No substrate inhibition
μ_{max} = 1.0 h^{-1}
k_s = 0.1 g l^{-1}
Y = 0.5
S_0 = 2.0 g l^{-1}
Parameters are valid also for b and c

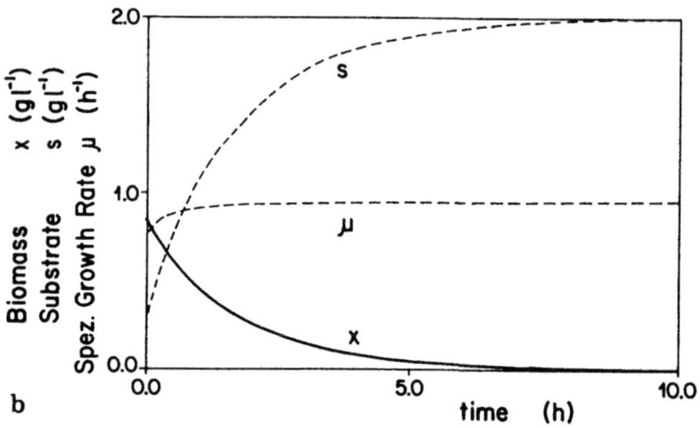

Fig. 2b. Simulation experiment as in Fig. 2a: Shift of dilution rate from 0.75 h^{-1} to 1.5 h^{-1}

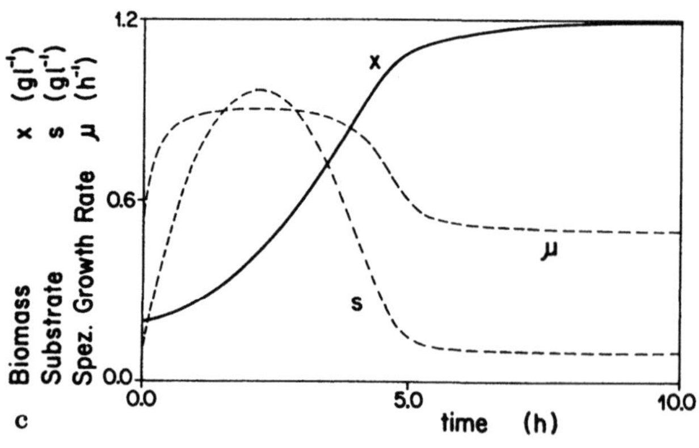

Fig. 2c. Simulation experiment similar to Fig. 2a: Shift of substrate concentration so from 0.5 to 2.5 g l^{-1}

Typical alterations in the x, s or r_x plots in function of D also become visible at various concentrations of the substrate s_0 (Fig. 1b). Again, the Monod equation is taken as a calculation basis. Washout decreases sharply at reduced values for D when the substrate becomes diluted, and the shapes of the curves in both Fig. 1a and b change their forms in a very characteristic manner.

In a similar way, *shift experiments* can also be of help in identifying controlling parameters. The curves of the plots for s and μ, respectively, in function of time (t) show that after shifting, D or s or other parameters are dependent upon chemical parameters (see Fig. 2a–c). For example, if the concentration of the limiting substrate is shifted (at constant dilution rates), a typical transient diagram will emerge. The diagrams of Fig. 2a–c, however, are obtained only under the provision that there is no substrate inhibition. Otherwise, washout will occur if $\mu < D$ at the new substrate concentration). Altogether, the plots in function of time of x, s, and μ during a shift of the dilution rate appear quite typical for a transient state based on the assumption given in the respective legends of the figures.

Finally, *pulse methods* are of equal value in determining decisive parameters of growth (Fig. 3a and b). The ideal form of a pulse experiment should be a small volume added for an infinitesimally small time period. The result would then be a sudden increase in the concentration of the substance pulsed, without change of D, provided that there was no uptake of the substance by the organism. Depending upon the residence time distribution of the mixing system, the original level of concentration

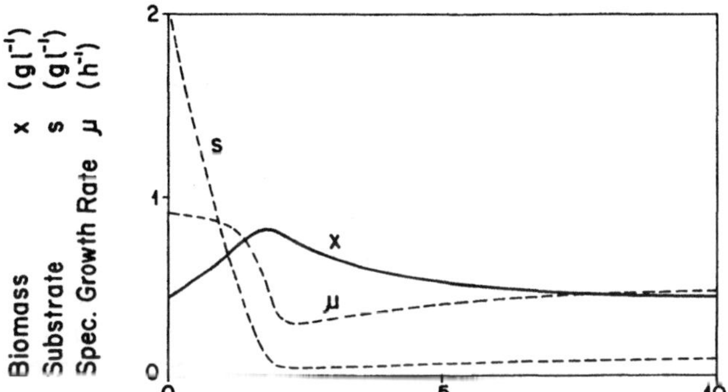

a

Fig. 3a. Pulse experiment: Limiting substrate. Pulse concentration = $2.0\ g\ l^{-1}$.
μ_{max} = $1.0\ h^{-1}$
k_s = $0.10\ g\ l^{-1}$
D = $0.50\ h^{-1}$
Y = 0.500
S_0 = $1.0\ g\ l^{-1}$

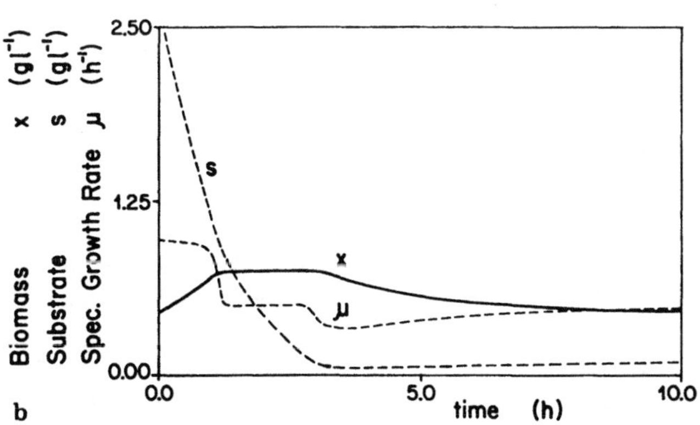

b

Fig. 3b. Pulse experiment at excessive pulse concentration = $2.5\ g\ l^{-1}$. Pulsed substance is not the only limiting medium component. Other parameters as in Fig. 3a

would be reached after the pulse. The response of the growth system, without any inhibition, would be as follows:

Case 1: The pulsed substance is the limiting medium component (Fig. 3a)

- μ increases immediately and decreases later (steady-state value prior to pulse not shown in the figures a and b)
- The cell concentration increases simultaneously according to the available substrate. It decreases after a maximum reaching the steady state asymtotically
- Decrease of the substrate concentration(s) to a level lower than the steady state value is typical.

Case 2: The substance pulsed is not the only limiting component (Fig. 3b) = double limitation

- Theoretically, μ cannot be maintained at the maximum despite the excess of substrate. The intermediate plateau in the curves of μ (also of x) reflects the limitation of a second (unknown) substrate. Under actual experimental conditions, however, small experimental oscillations may normally occur as an indifferent response to the pulse.

A large number of modifications (qualitative and quantitative) of the pulse methods are available to demonstrate the effects of single parameters. They represent, therefore, the most efficient way to evaluate the effects of actual or potential medium components of metabolic products.

1.4 Application

The experiments described in this section may give some ideas of how results of growth investigations can be evaluated. Continuous culture methods appear to be most valuable and of high efficiency, providing useful data in simulation work. Proper layout of pulse experiments, for example, can help researchers to choose useful concentrations for substrates and check for absence or presence of inhibitors and hidden limitations. Also, extremely valuable information about the strength of chemical parameters can emerge. In other words, chemostat processes, in all modifications, are extremely efficient and valuable tools for determining whether or not an observed phenomena is governed primarily by physical and chemical parameters.

In an overall view, the entire approach for the proper identification of all factors influencing growth behavior or product formation will remain empirical. However, the development of some rules for a more systematic procedure in control investigations is now possible. This will be shown in Sect. 3.

2 Physical Parameters

2.1 Physical Containments for Bioprocesses

2.1.1 Agitation and Aeration

Processes that make use of living cells as biocatalysts need adequate containments as efficient hardware components of small scale (research) and large scale (production) equipment. The proper layout of each containment is prerequisite for maintaining an optimal environment for the cells involved. Submerged culture systems are by far the most important methods of biotechnology, irrespective of the increasing relevance of immobilization or solid state procedures developed over the last one or two decades. However, because vessels used for immobilized systems may differ substantially from the agitated vessel, only reactor vessels for submerged cultures will be considered in the following.

Furthermore, emphasis will be placed on aerobic rather than anaerobic processes due to their industrial importance. Also, principles of mass transfer from gas to liquid and eventually to solid (cells) phases are valid for all types of gases involved in biology, and many of the results developed in Sects. 2 and 3 will be of value for those interested in anaerobiosis.

It is the aim of the following to discuss some aspects of fluid dynamics and mass transfer relevant to the application of agitation devices for bioprocesses. The trend of developing new reactor designs to meet requirements of energy-saving agitation and aeration hardware has become apparent. In view of the numerous proposals for new designs that allegedly improve the efficiency, operation, and economy of any bioprocess (all at low energy and investment costs), a critical review of the existing evaluation methods is presented. Finally, three new designs of bioreactors are described and characterized by new data that make use of biotest methods suitable for reactor evaluation.

Lowering the energy input for a mixing device cannot represent the final answer. In most cases, reaction mixtures consist of rather diluted solutions with only a small percentage of product in water. Consequently, product recovery — at least in its first steps — becomes a matter of large amounts of water handling, which requires energy. Optimization of a process must therefore include the costs for downstream processing, which may be (specifically) lower for highly concentrated solutions.

Demand for high productivity reactor types also comes from the fact that in plants for large scale production, such as for single cell protein, reactor volumes are reaching their feasible limits. Today, SCP is produced by ICI in Billingham, UK, in a 2300 m^3 vessel of 60 m in height and 7 m in diameter. Sewage is treated by Bayer AG in a 20,000 m^3 containment which has a height of 30 m und a diameter of 26 m. The enormous power input of 7 MW required for aeration demonstrates that mechanical agitation in the classical way is no longer feasible when very large plants are envisioned.

Obviously, reactor concepts that allow for high productivity are necessary. This goal, however, can only be reached if the most efficient devices for agitation/aeration — together with carefully designed media — are used to activate the full synthetic power of the cell.

Agitation and aeration are important aspects for the improvement of a bioprocess. The purposes of *agitation* are:

- to suspend the cells uniformly throughout the working volume in order to prevent sedimentation or flotation
- to minimize temperature and concentration gradients in the bulk of liquid
- to maintain high mass transfer rates by the uniform distribution of all reactants.

The mass transfer capacity of an agitation system is usually evaluated using the k_La values for oxygen. As oxygen is of poor solubility in aqueous solutions (< 10 ppm), its supply to the cell can become critical in a rather short time (< 30 s). A vast amount of literature on agitation and aeration is available today in which the earlier research concentrates more heavily on reactors for chemical processes. Since the fifties, when the antibiotic industry was growing substantially, bioprocesses were also included in the classical works on mixing [1-5]. Schuegerl [5], in his comprehensive review of bioreactors, lists the various systems for agitation (stirrer, vessels, etc.) and gives a thorough evaluation of their various designs. He also shows that data taken from model solutions are of no use for the potential optimization of a particular bioprocess and that valid comparisons of various agitation/aeration systems can be made only with data from bioprocesses that use identical strains and media. Any data found on rheology, k_La, or mixing effects must be checked for application to actual process mixtures.

Agitation studies have been based primarily on k_La studies of this type, placing much emphasis on the development of the design of the stirrer. Only more recently have studies on fluid dynamics in reactor vessels suitable for bioprocesses been done. An example is given in the fine review by Taguchi [6].

The energy demand for agitation has been at the center of all efforts for the development of agitation devices. The major characteristics of various containments, such as fluid and bubble movement affected by paddle, propeller, FBT or disk stirrers, have been collected by Brauer [7]. Basically,

$$\mathrm{Ne} = (\mathrm{Re};\ \mathrm{Ga},\ \text{geometrical parameters})$$

where

$$\mathrm{Ne} = \frac{N}{d_r^5 n^3 \varrho} \quad \text{(Newton number)}$$

$$\mathrm{Re} = \frac{n d_r^2 \varrho}{\eta} \quad \text{(Reynolds number)}$$

$$\mathrm{Ga} = \frac{d_r^3 g \varrho^2}{\eta^2} \quad \text{(Galileo number)}$$

Ne — Re relationship (non aerated, non baffled vessels):

laminar flow: $0 \leqq \mathrm{Re} \leqq 10^1$

$$\mathrm{Ne} \sim \mathrm{Re}^{-1}$$

transition from laminar to turbulent flow: $10^1 \lesssim \mathrm{Re} \lesssim 10^2$

$$\mathrm{Ne} \sim \mathrm{Re}\ 10^2$$

turbulent flow: $10^2 \lesssim \mathrm{Re} \lesssim \sim$

$$\mathrm{Ne} \sim \mathrm{Re}^{-0.28}$$

The important relationship of energy uptake of aerated and non-aerated stirred vessels is found to be quite variable:

$$\mathrm{Ne_G}/\mathrm{Ne} = \begin{array}{l} 0.4 \text{ paddle, FBT} \\ 0.3 \text{ propeller} \\ 0.1 \text{ disk} \end{array}$$

Due to the enormous variability in the physical and chemical characteristics of a cell suspension, optimization of the hardware for agitation and aeration is difficult. Bioprocesses are becoming increasingly specialized in one way or the other. The search for inexpensive raw materials has introduced substrates that have particulate characteristics (lipophilic liquids, solids), resulting in difficult distribution. Improvement of media and strains is leading to faster growth with greater demands for oxygen. Instead of chemical methods for O_2 transfer studies [2,8–11], new strategies such as bubble diameter distributions will be necessary.

The classical concepts of oxygen transfer as shown by liquid film limitation at the gas-liquid interface of critical dissolved oxygen concentrations certainly require refinement. Attempts therefore have been made to replace the classical, continuously stirred tank reactor (STR), which is now widespread in the bioindustry. Due to its wide use for all kinds of processes, this system will be discussed briefly.

Agitation in the STR is provided by a single or multi-stage flat blade turbine (FBT) in combination with several baffles. The typical flow pattern is given in Fig. 4 and has been discussed by several authors. Although a variety of STR configurations have been developed in the past, the geometry of an old proposal (1953) has become more or less 'standard'. Height to diameter ratio of the vessel is commonly 3 to 5 and the number of FBTs is 2 or 3, depending upon the size of the vessel. This agitator produces excellent turbulence at the impeller tip. Liquid is recirculated from the wall above and below each flat blade resulting in well-mixed regions. Mass transfer and mixing times within these regions are quite good, whereas the exchange of fluid from region to region may become limiting, especially in viscous media [19].

A thorough treatment of the limitations of mixing by the interchange of material between FBT regions is given by Sinclair and Brown [12]. This model prediction explains the observation that in chemostat experiments actual washout (D_c) is different from

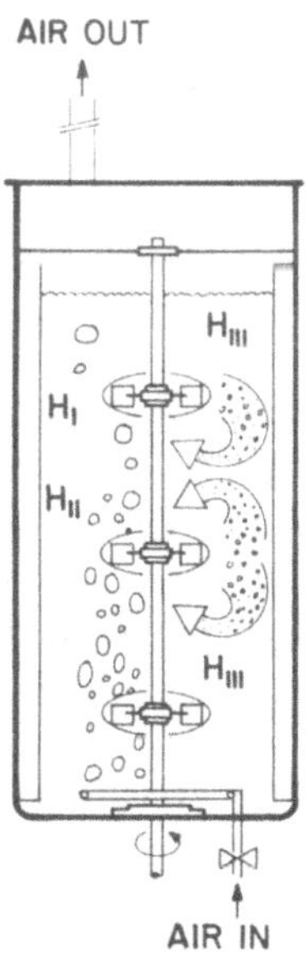

Fig. 4. Classical stirred tank reactor (STR) with 3-stage flat blade turbine (FBT) and baffles.
Typical ratio of height: diameter is usually about 3 (for other values see, for example, Finn [1]). Gas dispersion by sparger below the FBTs. Bubbles circulate in a region around each turbine as indicated by the arrows improving dispersion by repeated breakup. Typical fractions of gas-liquid mixture result as demonstrated in viscous media by Swallow et al. [19].
H_I Fractions around the FBT
H_{II} bubble diameter $d_B > 1$ mm
H_{III} bubble diameter $d_B < 1$ mm
Exchange of material between the zones of perfect mixing around the single turbines represents the limiting factor in the overall mass transfer of the mixture (see Sinclair and Brown [12]). This characteristic is one of the reasons for segregational effects of the content leading to the formation of a layer of cells and bubbles (and residual substrate) on top (see also Gschwend [20])

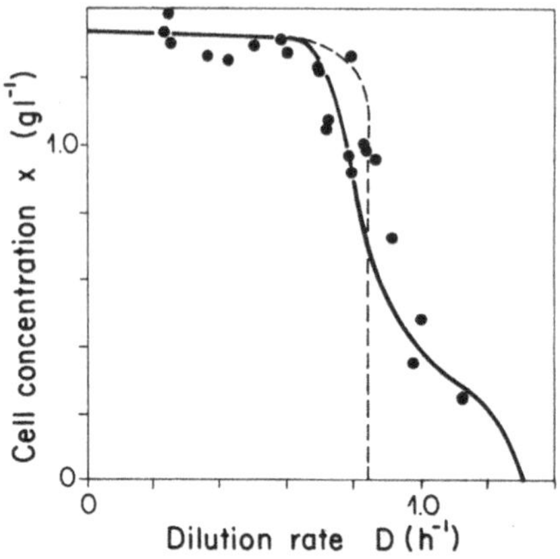

Fig. 5. Effect of non-ideal mixing properties of agitation device in continuous culture. Broken line indicates steady states in function of dilution rate D calculated according to Monod's equation. Full line calculated values for biomass based on a model developed by Sinclair and Brown [12] for the limited exchange of material between the mixing regions around the single FBTs. Single plots represent experimental data from *Escherichia coli.* (Herbert et al. [113]). The difference between experimental values and simulation according to Monod represents the extent of incomplete mixing

its value calculated according to Monod (see Fig. 5). This difference shows that inadequate mixing behavior may lead not only to erroneous interpretations of the phenomena observed, but also prevent any precise experimental work. Stability of the steady state in continuous cultivation is poor and possible oscillations in signals obtained from on-line measurements may be altered (e.g., Meier and Beyeler [13]).

The importance of mixing on washout and steady state performance in continuous cultivation has been known for quite some time (Fan et al. [14]). These authors investigated the effects from extreme micromixing (complete segregation) to maximum macromixing. Their results are of considerable interest allowing improvements of mixing devices based on demostat experiments.

2.1.2 Practical Problems of Operation

Concepts for bioreactor design are not easy to develop since the nature of the materials to be processed can vary considerably. As a fully optimized solution for each specific process is not possible, the versatility of bioreactor constructions is most important.

Current reactor designs follow various concepts. One of them is the postulate for minimizing power input P with the consequence that plants for high production goals may become too large. A second concept is to increase P/V in order to increase productivity. To do this, kinetics (μ) and/or concentration ($s_0 \rightarrow x$) or both must be enhanced. However, high medium concentrations may involve the drawback of limiting the solubility of the components, salting out effects, or other chemical side effects. Also, foam formation may disturb the process or special procedures for the sterilization step must be developed. The addition of antifoam agents, for example, can be especially disturbing. It may interfere with downstream processing for product recovery or waste treatment and represent an undesirable cost factor. Membrane assisted product recovery or production of surfactants are no longer processes permitting chemical antifoam abatement. Other strategies such as mechanical destruction of foam must be developed.

These few examples show exactly how bioprocesses are systems that are not always simple to handle. Aspects such as cell material, cell specific O_2 demand, its physical nature (viscosity, surface tension) and chemical nature (ion concentration, salt concentration) have already been indicated. These require a great adaptability of agitation and aeration devices in order to make a process feasible. It is therefore of merit that some of the more recent publications have begun to deal with the special properties of the mixtures involved, such as the rheology of mycelial culture fluids [15], microbial flocs formation and floculation [16], significance of microbial films [17], solubility of medium components [18], and influence of non-Newtonian fluid characteristics [19, 20]. Some examples of the drawbacks of incomplete mixing in continuous systems are given in Fig. 6. Wall growth, flotational effects or sedimentation can cause drastic artifacts in the apparent kinetic performance of an organism.

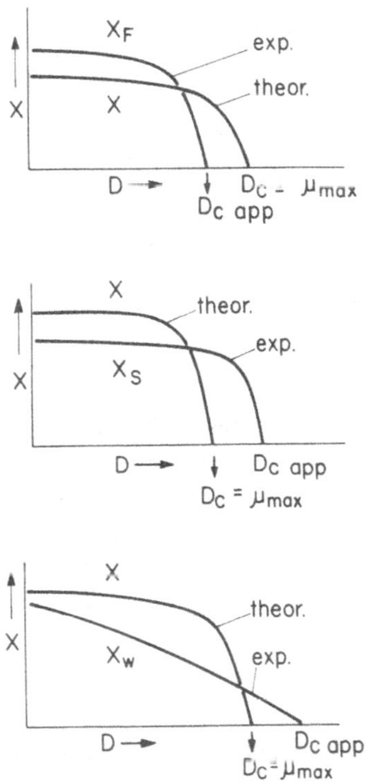

Fig. 6. Effects of inadequate mixing in continuous mixing. Artifactual data for washout or biomass concentration occur due to flotation (TOP: X_F), sedimentation (CENTER: X_s) or wall growth (BOTTOM: X_w) (after Humphrey) plots for experimental and theoretical values are given for biomass x

A perfect hardware system, consequently, must not only maintain high mass transfer rates and short mixing times, but also prevent concentration profiles, any effects of flotation/sedimentation, wall growth, overflow of foam, and last but not least, ensure reliable operation under sterile conditions. In many cases, these practical aspects can drastically reduce the economy of an otherwise perfect bioreactor.

A practical solution to the problem of the perfect mixing of a multi-phase (gas, liquid, solids) system is illustrated in Fig. 7. Air flow through an otherwise well-mixed volume (fraction 1 of the upper Figure) exerts segregational effects leading to the so-called 'foam' layer (fraction 2). Fraction 3 of the volume, generally known as 'headspace', is thought of as a means for the prevention of foam overflow. This function, however, is very often inefficient, particularly under heavy foam formation

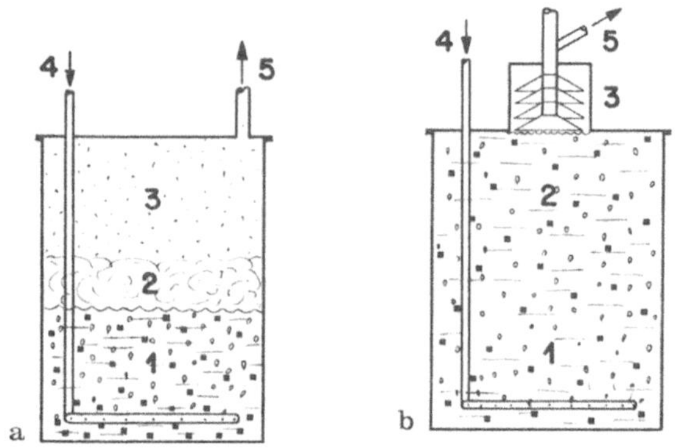

Fig. 7a and b. Perfect mixing of a multiphase system containing gas, liquid and solid phases is not simple to achieve. In classical devices (agitation elements are not depicted) with head space for "foam" expansion, flotation effects may occur. **a.** (1) mixture of liquid — gas bubbles o and solids (cells, substrate) o segregational effects may occur due to air flow (aeration) irrespective of the agitation system applied. The gas fraction included in this part of the volume is related to the physical properties of the reaction mixture. (2) "Foam" layer containing enriched cell content. Foam formation cannot be brought under control easily. Addition of antifoam agents interferes also with the properties of fraction one leading to uncontrollable gas hold-up; (3) Head space, containing cells on walls and exhaust gas enriched with CO_2; (4) air feed (5) exhaust gas; **b.** (1) Practical solution to the problem of how a multiphase system can be kept under controlled dispersion during agitation/aeration; (2) homogenuous distribution. Both fractions are identical; (3) mechanical phase separator (no chemical antifoam agents involved in foam treatment); (4) air feed (5) exhaust gas

conditions. It also provides poorly controllable conditions for ideal mixing (wall growth, remainders of medium and flow pattern due to changing medium level).

A possible practical solution for 'ideal' mixing of the phases involved is given in the lower picture of Fig. 7. Whatever the principle of agitation might be, segregation in a completely filled reaction volume cannot occur due to holdback of the liquid/solid phases by mechanical phase separation. Configurations as shown in Figs. 9 and 11 are examples of the application of this principle.

2.1.3 Conclusions

Theoretical and practical considerations indicate the need for new solutions in reactor designs. Principles for agitation and aeration have not changed much since the classical STR type with FBT agitation was adopted for antibiotics production. During the seventies, as the result of a programmatic study sponsored by the Federal Ministry of Science and Technology in Germany, several attempts were made, primarily in Europe, to develop real alternatives. This became effective after the so-called air-lift principle, e.g., agitation by a rising swarm of bubbles, was first proposed in the USA.

Approximately nine types of bioreactors represent more or less significant examples. They have been reviewed in several articles by Sittig [21], Schuegerl [22] and Schuegerl [5]. It becomes clear that each physical containment has to be adjusted to the needs of the actual working dispersion and not be 'optimized' for model solutions.

An entire array of more practical requirements must be fulfilled in reactor designs besides the classical aspects such as P/V, k_La, or at best gas holdup, especially for large units (production plant size). These aspects include configuration as such, specific surface of heat exchange, total energy input, total production costs (including downstream processing), variable costs and potential productivity. The rather low productivity of many bioprocesses has supported the idea of using low power input instead of intensifying investigations on metabolic control of growth and product formation by studying the effects of extracellular parameters.

In the following Sect. 2.2, a selection of some new principles of bioreactor configurations will be shown. This selection concentrates on two layouts using mechanical agitation and two using other types of agitation. The latter two were developed mainly to overcome the necessity of using high power input in very large sized units.

Two reviews by Schuegerl on the present situation in reactor design are probably the best basis for developing evaluation material [5,23]. His experimental work has concentrated on air-lift configurations (bubble column, tower reactors). Figure 8 shows the airlift principle and a selection of possible modifications. The basic characteristic is agitation by the air stream for aeration.

In Table 1, the differences in the fluid dynamics and mixing behavior of the classical STR and tower forms (TR) are given, as produced by this author. The advantages and drawbacks of both types are visible, and it is noteworthy that the TR forms fit primarily to the idea of low specific power input. Accordingly, productivity must be principally low as well. Schuegerl has investigated this configuration extensively [70-74], employing certain bioprocesses for the testing of performance. Values for the productivity of yeast cells growing on ethanol or glucose given in Table 3 show the effectiveness of a TR at low P/V.

In a format similar to that of Table 1, the fluid dynamics and mixing behavior of three other configurations are given in Table 2. It includes two loop forms of principally different geometry and agitation principles. The compact loop reactor (COLOR) is an internal loop (for systematics see Blenke [26]) with a propeller-type stirrer for agitation. The geometrical ratio of height to diameter is practically 1 in order to maintain a high frequency of recirculation. The second loop form has a tall geometry (height:diameter = 10; TLR) with jet agitation (JLR). The tower-like

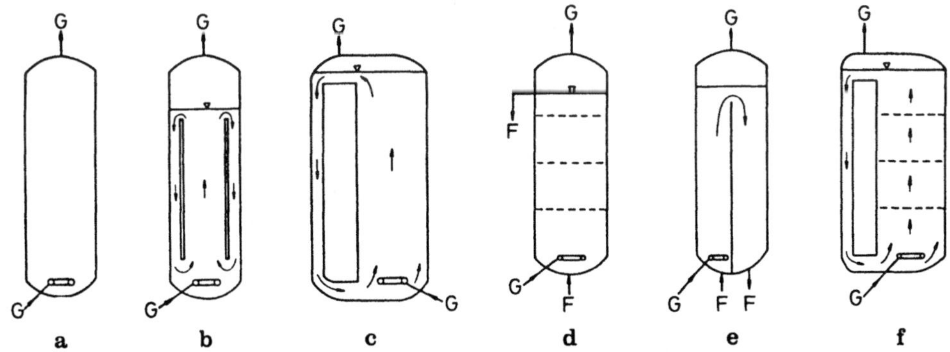

Fig. 8. Principle of air lift reactor (top left). Several possible modifications are given including draft tube (top center), plates (bottom left and right) and other elements for flow control. Agitation of liquid occurs solely by the air injected for aeration. The numerous modifications in air lift configurations show the necessity for the improvement of distribution in all types of built in elements like draft tubes, plates with holes etc. (F = liquid, G = gas) (after Schügerl [23,24])

Table 1. Characteristics of *continuous stirred tank reactor* (CSTR) vs. *tower reactor* (TR). (Schügerl [231])

Characteristic	CSTR	TR
	I. Fluid dynamics	
Agitation	Stirrer	Bubbles
Primary stream	Rotational	Upward directed
Secondary stream	Formation of eddies (laminar flow condition)	Formation of eddies (laminar flow conditions)
Transient to turbulent	Decay of eddies (superposition of several periodic agitations)	Decay of eddies (superposition of several periodic agitations)
Turbulence	Stochastic & periodic component	Purely stochastic
Energy dissipation rate	Non-uniform	Uniform
Gas dispersion	Repeated dispersion of bubbles	Redispersion of bubbles rare
P/V	High	Low
	II. Mixing behavior	
Dispersion (large volume level)	Good	Moderate to poor
Residence time distribution	Broad	Broad to narrow
Local dispersion	Good	Moderate to poor
Mixing time	Short	Long to average
Air and energy input	Non-coupled	Coupled
Distribution of concentration of the mixed phases (salts, nutrients, dissolved O_2)	Strongly coupled	Partially non-coupling tendency
Reaction models	With concentrated parameters	Partially with concentrated and distributed parameters, respectively
Potential for reaction optimization	Very low	Moderate to good

layout provides excellent use of oxygen from air bubbles due to the long residence time. Agitation in this configuration is not restricted to jet-injected (withdrawn) working liquid, as agitation with the stirrer may also be possible. However, this latter version has found less application so far than the jet. The flow principle of Table 3 is in some ways related with the loop form as it basically represents a geometrically ideal form for the forced circulation of a fluid along a closed way. The annular form has been named TORUS and is equipped with mechanical agitation.

All three configurations of Table 2 are shown in Figs. 9–11. They will be described in more detail in Sect. 2.2, which will also include some characteristic data from growth experiments.

2.2 Reactor Design for Bioprocesses

2.2.1 New Bioreactors

There are an impressive number of reports dealing with new proposals for bioreactor configurations. From this it becomes clear that there is a need to adjust the construction to meet the needs of each particular process. In the seventies, the major problem was the need for large plants for the production of single cell protein (SCP), which required agitation devices for working volumes of several hundred cubic meters. Unfortunately, it seems that little attempt was made in those days to evaluate the designs with actual bioprocesses. Most reports used the classical approach of hydrodynamic characterization. Data from optimization work with actual fluids with active cells are therefore rare.

In this Section, a small selection of new bioreactors will be discussed. Their main features are given in Table 4. It is noteworthy that comparable data of cell productivity from use of the same biotest system are available. Furthermore, data from more than one size of some reactors are available. The data given in Table 4 refer to the following reactor types.

Form	Name	Tables
Loop — short	Compact loop reactor	COLOR
— long	Tall loop reactor	TLR
Annular	TORUS	TORUS
Tower	Air lift bubble column	TR
Classical STR	Continuous stirred tank reactor with flat blade turbine	FBT

The characterization of a physical containment is reliable only when the growth pattern and its metabolic control are known precisely. For this, a well elaborated medium that provides a *defined* limitation under all circumstances of growth is required. Although this sounds simple, it is actually not possible for all types of microbes and products. Complex growth patterns as, for example, encountered in antibiotic producers are not suitable. The genetic instability of the strains involved and insufficient knowledge of the relationships between medium and growth or product formation make these poor candidates for a definable test system. Thus, organisms with a simple product spectrum such as cell mass, CO_2, and H_2O with excellent growth in a completely defined medium are much better suited for this purpose. For checking the mass transfer potential of an agitation device, a strain with only a high biosynthetic capacity is required. Methods for systematic medium design and for development of biotest systems suitable for bioreactor testing are outlined in Sect. 3.

Table 4 shows a condensed form of the main features of four selected designs different from an FBT equipped vessel. Some comments on the background of their construction and performance may be useful.

2.2.1.1 COLOR

This design is based on earlier proposals of the sixties (Fiechter [27]) designed to overcome inadequate mixing effects observed in yeast cultivation. An optimized configuration is shown in Fig. 9. The segregation of cells from the liquid phase (flotation) that occurs in a classical FBT bench-scale model was overcome by increasing the so-called macromixing component ('blending') of the agitation. The circulation

Table 2. Characteristics of short loop reactor (SLR) e.g. COLOR, tall loop reactor (TLR) e.g. JLR, and annular form reactor (TORUS)

Characteristic	SLR	TLR	TORUS
	I. Fluid dynamics		
Agitation	Mechanical (propeller)	Hydrodynamic (jet) or mechanical	Mechanical (propeller)
Primary stream	Circular (vertical)	Circular (vertical)	Circular (horizontal)
Secondary stream			
laminar flow	Formation of eddies	Formation of eddies	Formation of eddies
transient to turbulent	Decay of eddies (superposition of several periodic agitations)	Decay of eddies (superposition of several periodic agitations)	Decay of eddies (superposition of several periodic agitations)
turbulent	Stochastic & periodic component	Stochastic	Stochastic & periodic component
Energy dissipation rate	Uniform	Non-uniform	Uniform to non-uniform
Gas dispersion	Periodic dispersion of bubbles	Local dispersion of bubbles	Periodic dispersion of bubbles
P/V	Moderate to low	Low	Moderate to low
	II. Mixing behavior		
Dispersion	Very good	Moderate to poor	Very good
Residence time distribution	Broad	Broad to narrow	Broad
Local dispersion	Good	Moderate to poor	Good
Mixing time	Short	Long to average	Short
Air and energy input	Non-coupled	Jet-agitation: coupled mechanical: non-coupled	Coupled or non-coupled
Distribution of concentration of the mixed phases (salts, nutrients, dissolved O_2)	Strongly coupled	Strong tendency for uncoupling	Strongly coupled
Reaction models	With concentrated parameters	Partially concentrated and partially distributed parameters	Partially concentrated and partially distributed parameters
Potential for reaction optimization	Good	Moderate to good	Good

	III. Application aspects		
Control of flow pattern (at high gas content)	Possible	Hard to achieve	Possible
Foam control	Mechanically or chemically	Chemically	Mechanically or chemically
	IV. References		
	Fiechter [56]	Blenke [26]	Läderach [30] Gschwend [20]
Type of Table 4	COLOR	JLR, PLR	TORUS

Table 3. Productivity (r_x) and q_{O_2} of *Hansenula polymorpha* in various batch experiments in a tower reactor (TR) with external loop [55]

Substrate	Medium limitation	Productivity r_x $g\,l^{-1}\,h^{-1}$	Oxygen transfer rate OTR $g\,l^{-1}\,h^{-1}$
Ethanol	O_2	2.5	4.7
	ethanol	0.95	1.7
Glucose	O_2	3.3	2.8
	glucose	1.7	1.4

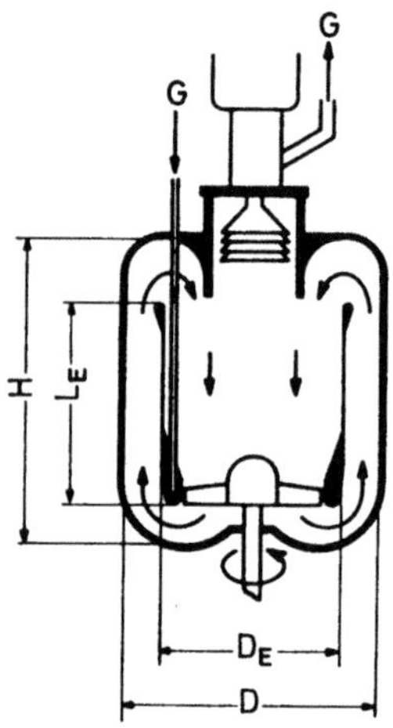

	CFR 50	CFR 550
H	470mm	1650mm
L_E	290mm	1300mm
D	400mm	700mm
D_E	280mm	500mm
V	50 l	550 l
H/D	1.2	2.4

Fig. 9. Compact loop reactor (COLOR). Compact forms of internal loop configurations are obtained with a small ratio of height to diameter ≡ 1–2 giving rise to short mixing times (See also Figs. 12 and 13). Such a principle is an adequate form according to Fig. 8. It can be a operated in a completely filled state avoiding any inhomogeneity within the vessel. The flow pattern can be kept under perfect control regardless of the actual gas hold-up. *No addition of antifoam agent needed*

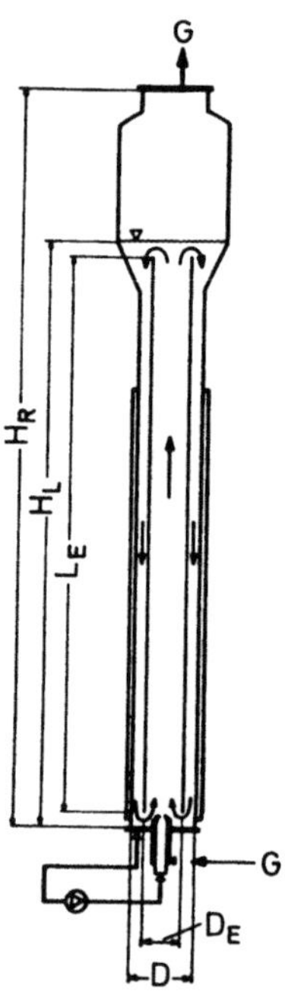

H_R	6580mm
H_L	5320mm
L_E	5010mm
D	575mm
D_E	345mm
V_L	1.4 m^3

Fig. 10. Jet loop reactor (JLR) (by Blenke [26]). The tall form (height: diameter ~ 10) provides an extended residence time. Consequently, the air at exit is well depleted of oxygen. A rather large head-space is needed to prevent overflow of foam

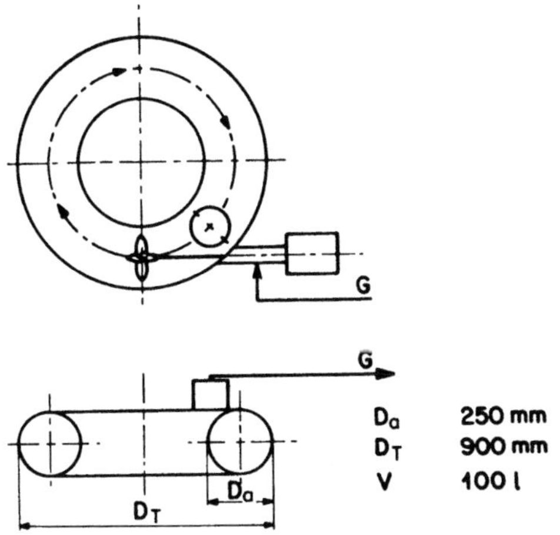

Fig. 11. Annular form (TORUS) with mechanical drive (see Laederach [30]). The mixture is circulated along a given pathway. The resulting flow pattern is therefore well under control irrespective of the physical nature of the liquid. Short mixing times result due to the rather high circulation frequency. G = gas. Phase separation at gas exit by mechanical separator. *No addition of antifoam agent needed*

overflow pattern obtained by combining a central draft tube with a ship type propeller yielded an excellent mixing performance but did not prevent further flotation when heavily foaming media such as molasses solutions were used. However, using mechanical means for foam breakdown on the top and appropriate adaptation of the geometrical form resulted in a device with good mass transfer properties and perfect control of the flow pattern. This vessel can be operated in a completely filled state. The addition of any antifoam agents is unnecessary. By adjusting the actual filling degree of the medium used to the separation capacity of the foam breaker, segregation of the dispersion or outflow of 'foam' is impossible. The permanent 'washing' of all internal surfaces (including inserted sensors) prevents growth of cells along the walls. A 50 l COLOR reactor was later characterized both hydrodynamically and biologically [28] and these data were used for optimization [29]. Figure 12 gives evidence of the short mixing times of 2–3 s that are obtained with power input of 2–3 Wl^{-1} (water) or 8 s at 10 Wl^{-1} with 2% carboxymethyl cellulose mixtures. The agitation principle applied in this design allows one to adjust the mixing time using only one parameter: propeller speed rpm. Circulation time T_G is a linear function of mixing time T_{mix}, as shown in Fig. 13. Actually, high productivities of more than 10 $g\,l\,h^{-1}$ of cells were obtained in later experiments in this 50 l reactor and in the larger units, as shown in Table 4. The use of this type of reactor, as well as smaller bench-scale models has revealed its usefulness and great adaptability to all types of bioprocesses, including bacteria, yeasts, and microfungal systems.

2.2.1.2 Tall Loop Reactor, esp. jet loop (JLR)

This proposal of a low power consuming containment has been elaborated by Blenke. A full systematics of this tall form of loop reactor has now appeared [26]. Agitation may be affected hydrodynamically as well as mechanically. The jet form of Table 4 and Fig. 10 was developed in view of large plants for SCP production and was designed for low specific power input but high usage of oxygen via bubbles. This latter feature is primarily possible because of the long residence time of the bubbles resulting from a long travel path and slow movement. This reactor type, with all of its modifications,

Table 4. Physical characteristics of a selection of bioreactor designs

Configuration	Loop forms			Air-lift [33]			Classical STR
	Short (SLR)	Tall (TLR)	Annular	Column	Bubble column	Column	FBT baffled
Common name	COLOR	JLR	TORUS		Porousplate	Ejector injector	
Geometrical characteristics	H/D = 1 (max. 3)	H/D ≧ 10	Circular cross-section		H/D ≧ 10		H/D = 3 with baffles 1/10 of d_R
Fig. No.	9	10	11	8			
Ref. No.	28, 29, 54, 56–64	26, 65–60	20, 30, 31	24, 25, 69–74	24, 126	24	80
Agitation	M	P(H), M	M	P	P	P(H)	M
V_{tot}, L^-	50 220 390 550	2300	20 100 1000	—	—	—	30 300 3000
$V_{eff}:V_{tot}$	>0.8	~0.6	~0.8		~0.7		~0.7
P/V, kWm^{-3} (actually investigated)	23 16 12 10	1	6	—	≦1		~5
k_La, Ref. see	—	—	—	$\alpha p^{0.75}$ 79, 82	— 15, 17, 19, 26, 27	—	$\alpha p^{0.8}$ 14–17, 79

OTR, g L^{-1} h^{-1}	6 5 5 4	2	4	—	—	—	4 4 3
t_{mix}, s Ref.	—	70	60	— 22–25)	— 15, 17, 19, 26, 27)	—	—
Productivity g Biomass L^{-1} h^{-1}	11 9 12 8	5	7	—	—	—	10 10 8

SLR short loop reactor (Swiss Federal Institute of Technology, Zürich);
JLR jet loop reactor;
PLR propeller loop reactor;
FBT flat blade turbine;
M mechanical;
H hydraulic;
P pneumatic

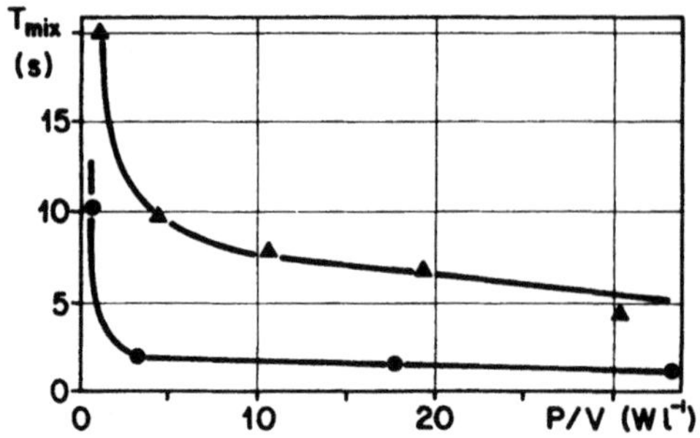

Fig. 12

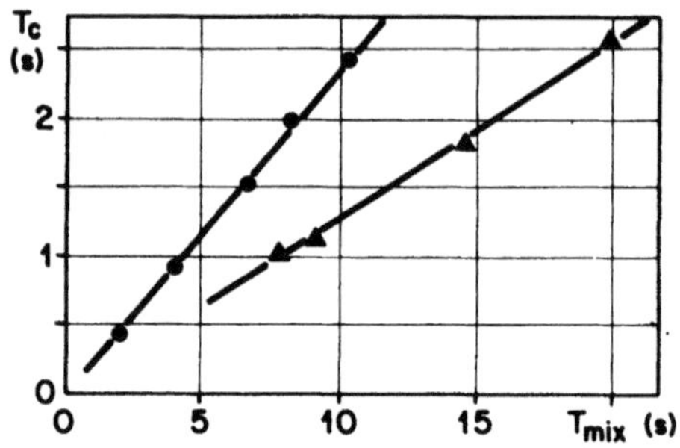

Fig. 13

Fig. 12. Compact loop reactor (COLOR) 50 l configuration (Karrer [29]). Mixing time (T_{mix}) vs. power input P/V obtained from two different model liquids:
—●—●—●— water;
—▲—▲—▲— carboxymethylcellulose, 2%.
Total volume of the vessel = 50 l; actual volume of liquid = 46.9 l. Very short mixing times can be maintained in this type of reactor at low energy input, which is actually about 10 × less than in the present optimized configurations

Fig. 13. Same configuration as in Fig. 12 showing circulation times T_c vs. circulation time T_{mix}.
—●—●—●— water;
—▲—▲—▲— carboxymethylcellulose, 2%.
The perfect distribution of the reaction mixture and the short mixing times in the compact loop form are achieved by fast recirculation of the vessel content (Karrer [29])

has been well characterized hydrodynamically (for references, see Blenke [26]). First biotests reveal that productivity is, according to its low power input, rather modest. Sterility problems do not appear in the vessel itself, but in the external loop where appropriate types of pumps must be selected. An additional cooling capacity can be easily created using heat exchanging devices included in the external loop. To avoid foam overflow, a rather large headspace at the top is necessary. In this part of the vessel, control over consistency of the flow pattern is poor if foam formation becomes substantial.

2.2.1.3 TORUS

The circulation of a fluid by a propeller in an annular geometry represents a simple principle for improving mixing on a macro level. The TORUS as shown in Table 4 was proposed by Widmer (ETH-Zurich) to overcome the drawbacks of limiting overall mixing. Its hydrodynamic properties have been described extensively [30,31] and characterized with biotesting [20,32]. The design shown in Fig. 11 has excellent properties for forced circulation in Newtonian and non-Newtonian fluids (Fig. 14); for elevated oxygen uptake rates, as compared to the STR, in yeast and mycelial systems, and for much higher productivities of yeast, as compared to the FBT system in both Newtonian and non-Newtonian fluids (Fig. 15a and b). It is obvious that this annular principle, similar to the short loop, exerts strong control on the hydrodynamic flow and can also be easily adapted for appropriate foam control (see Laederach [30]). As Gschwend [20] has shown, the volume fraction active in mass transfer is rather restricted (20–25%) at the relatively low power input values used

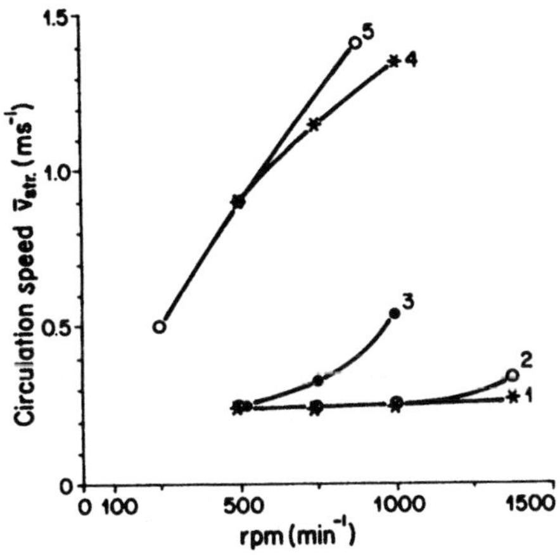

Fig. 14. TORUS 100 l (Gschwend [20]). *Liquid speed vs. rpm. Parameter is nature of liquid processed*

No.	Fluid	F_G	Author
1	D 2.5% + 0.5% xanthan	5.8	Gschwend [20]
2	D 2.5% + 0.5% xanthan	2.9	Gschwend [20]
3	D 2.5% + 0.5% xanthan	0	Gschwend [20]
4	water	5.8	Laederach [30]
5	water	0	Laederach [30]

F_G = gas flow rate

in his experiments. However, this limitation from full exploitation of the total volume may be reduced by using a higher energy input, installing more effective air injection to better separate agitation and aeration, or adjusting the baffling to improve turbulence on the way down to recirculation.

The three systems described thus far have an important feature in common in that emphasis has been placed on achieving fast and controlled agitation of the fluid. In the first case (COLOR) and the third case (TORUS), this was realized due to the forced circulation of the fluid along a defined pathway. This principle allows for best adjustment to the specifics of a single biosystem, resulting in the process optimization desired.

2.2.1.4 Air-lift Reactors

The air-lift principle was probably first proposed by Wang and Humphrey [33]. Since then, a great deal of experimental work has been done, as can be seen by the reference list of Table 4. The air-lift reactors are characterized by their agitation principle, which makes use of air to accelerate fluid movement in a suitable manner. This is achieved with a whole array of construction forms, including bubble columns and all kinds of towers with injecters, ejecters, or porous plates for air dispersion. Besides fluid dynamic investigations (for review, see [5, 23]), biotesting has been performed using *Escherichia coli* [74] and yeasts (see Table 3).

On the whole, air-lift reactors perform well, showing appreciable productivity at low power input rates and requiring simple operation. As mechanical aeration

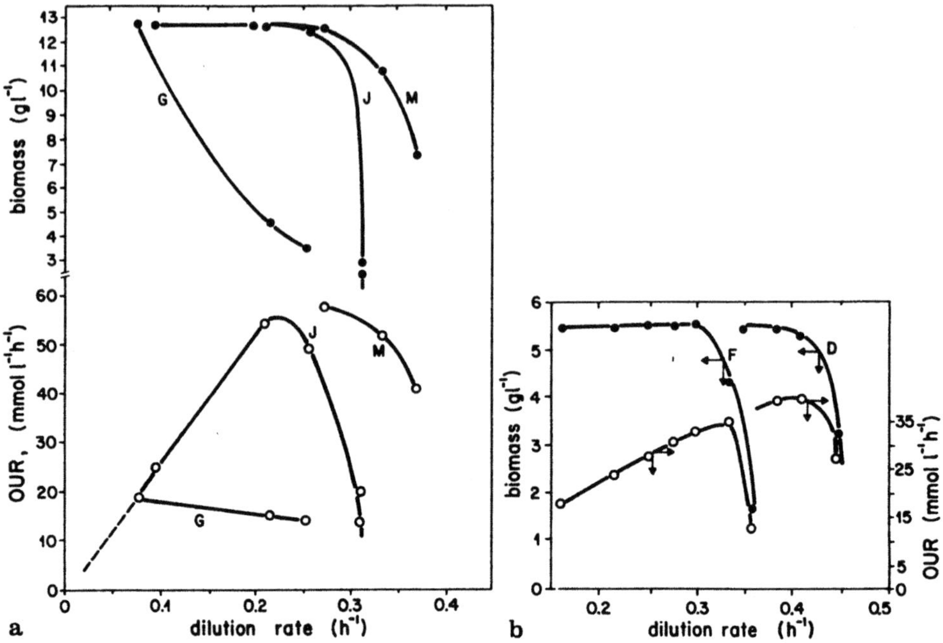

Fig. 15a. Performance of TORUS 100 l in biotesting with non-Newtonian fluids. Continuous cultivation, *Trichosporon cutaneum* (Gschwend [20])

Plot	rpm	P/V W l⁻¹	Medium	% O_2 of supplied air
G	750	1.2	D 2.5% + xanthan 0.5%	21
J	1375	3.6	D 2.5% + xanthan 0.5%	21
M	1250	3.3	D 2.5%	21

The limitations in mass transfer by low power input and increased viscosity (G, J) of the fluid are well demonstrated by these experiments; **b.** Biotesting of FBT 5 l (Gschwend [20]). Biomass and oxygen uptake rate (OUR) in function of D (continuous cultivation of *Trichosporon cutaneum*) Medium D 1.0% + xanthan 0.5%.

Plot	rpm	P/V W l⁻¹	
D	1000	5	3 stirrers
F	1000	5	1 stirrer (upper 2 removed)

The drastic reduction of biomass formation in presence of xanthan (o–o–) is a characteristic of the FBT-system due to the sharp reduction of turbulence in the bulk of the fluid

is not used, operation and maintenance, including sterility control, appear to be simple and inexpensive. This gives a good basis for scaling up to very large production plants. Increase of productivity, however, cannot be achieved substantially unless new devices for power input and for proper control against flotational effects can be set up. Nevertheless, it should be kept in mind that the largest units presently in operation are based in one way or the other on the air-lift principle.

2.2.2 Evaluation of the Physical Parameters

Table 4 has been composed for the presentation of current trends in bioreactor development. Together with Tables 1 and 2 on fluid dynamic properties, these data show that the classical FBT-agitation principle will be abandoned provided that real alternatives are available. Power input, although still important, will no longer be the only basis for design as more differentiated concepts are developed. Designs with the potential for process optimization will be important, and it is most likely that very large units will differ substantially from the hardware for bench or small pilot-plant scale. In this second category, perfect mixing with short mixing times, effective control against segregational effects without addition of antifoam agents, and good potential for mass transfer in actual process fluids will probably remain in the forefront of construction considerations. Needless to say, exact mixing is an important requirement for the proper, on-line measurement of process parameters. This holds especially true for the bench scale units used in research in which process control in the future will require extended automatic analysis for extensive direct digital control (DDC).

2.3 Biological Test Systems

Any testing and evaluation of hardware for aeration/agitation will be reliable only if data from specific processes are available. This implies the involvement of microbes with well known metabolic performance and of media with known chemical composition and physical properties. Only on such a basis will comparisons of various designs be possible.

For testing various features such as mass transfer, gas holdup, mixing time, segregational tendencies, foam abatement, capabilities for treatment of mycelia or susceptibility of cells to shear stress, a suitable biosystem has to be selected. Table 14 gives some possible combinations of strains and substrates for this purpose. In some cases such as molasses media, the system lacks the necessary definition of chemical composition due to the variable components found in molasses. However, molasses media are well suited for testing the foam control properties of a device such as a mechanical foam breaker. For testing other properties such as mass transfer capability, strains with a high oxygen demand are best.

When using Table 5, special care must be taken in regard to the composition of media. Not every reference contains a recipe suitable for the purpose of biotesting as the actual limiting components may be unknown or some other supplement may exert an unwanted bioregulatory effect. Proper medium design as outlined in Sect. 3 is recommended before time-consuming work with reactor testing is done. Needless to say, chemostat methods for medium design are of utmost importance.

The proper adjustment of a selected biosystem to the needs of an experimental approach also must be made on the level of medium strength. In Fig. 16, the time course of a batch cultivation of *Trichosporon cutaneum* is given. The resulting growth pattern evolved by this glucose-insensitive yeast includes an exponential phase plus a linear phase, indicating the onset of O_2 limitation. This set of data for typical limitation is obtained after proper adjustment of the medium concentration. It must

Table 5. Overall analysis of dry yeast. Composition of the cell is important in mass balance studies as well as in nutritional reflections

a) Main atom content of yeast		
Atom	*Sacch. cerevisiae* (White [89]) %	*Candida* (Sperber [90]) %
C	47.0	45.94
H	6.0	6.72
N	8.5	7.31
O	32.5	32.08
Ash	6.0	7.75
b) Main components of yeast (Altmann and Dittmer [114])		
	Brewer's yeast %	*Candida* %
Dry matter	93.0	93.0
Ash	6.4	7.8
Crude fiber	3.0	2.0
Ether extract	1.1	2.5
Protein		
N × 6.25	44.6	48.3
Digestible	38.4	41.5
	Major minerals	
Ca	0.13	0.57
Fe	0.01	0.01
Mg	0.23	0.13
P	1.43	1.68
K	1.72	1.88
Na	0.07	0.01
	Trace elements (mg kg^{-1})	
Co	0.2	—
Cu	33.0	13.4
Mn	5.7	38.7
Zn	38.7	99.2
	Vitamins (mg kg^{-1})	
Thiamine	91.7	6.2
	Brewer's yeast	*Candida*
Riboflavine	35.0	44.4
Nicotinic acid	447.5	500.3
B_6	43.3	29.5
Biotin	—	1.1
Pantothenic acid	109.8	82.9
Folic acid	9.7	23.3
Choline	3885.2	2910.6

be mentioned that special care in proper medium design and adjustment of its concentration is necessary when testing is done, for economic reasons, in batch cultures in large units instead of continuous cultures.

It is interesting to note that the bioengineering reports published in the past contain an appreciable amount of data from biological experiments related to the potential of mixing hardware. However, only a few of them have demonstrated the actual limiting parameters, which represent the controlling factor of the system.

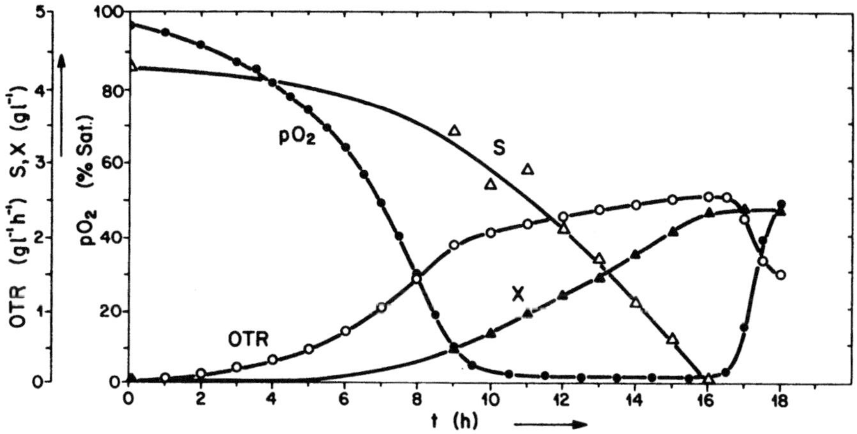

Fig. 16. Culture of *Trichosporon cutaneum* in a defined medium D used as a biotest system for bioreactor evaluation [38]. An appropriate medium concentration for yielding sequentially linear growth behavior after an exponential phase is chosen for visualization of the mass transfer potential of the agitation/aeration device

3 Chemical Parameters

3.1 Kinetic and Mass Balance Considerations

Microbial growth can be best described by assessing kinetics and mass balances. The kinetics of proliferation of cells or the release of products formed correlates strongly to the activity of the enzymes involved in the metabolic turnover of substrate, the formation of intermediate compounds, the synthesis of cell components and the secretion of products.

Monod was the first to demonstrate that the kinetics of growth in his bacterial systems very much resembled the Michaelis-Menten equation for enzymatic substrate conversion to product in a buffer system. Monod was finally able to present experimental data which fit exactly to the mathematical relationship for enzyme kinetics in non-inhibited systems

$$\mu = \mu_{max} \frac{s}{K_s + s} \qquad \text{Monod [34]} \qquad (1)$$

It follows that Monod was actually the first to recognize the great variability possible in specific growth rate. Its maximum value can be reached only under the condition of excessive medium components and it depends on physical parameters, as discussed in Sect. 2. However, in Eq. (1), μ_{max} functions as a constant and its quantitative value therefore stands for degree of optimization of all parameters related to growth. In contrast, μ as a variable depends only on the availability of the medium. The Monod equation consequently deals with a large number of enzymes and leads to a description of their overall, actual activity.

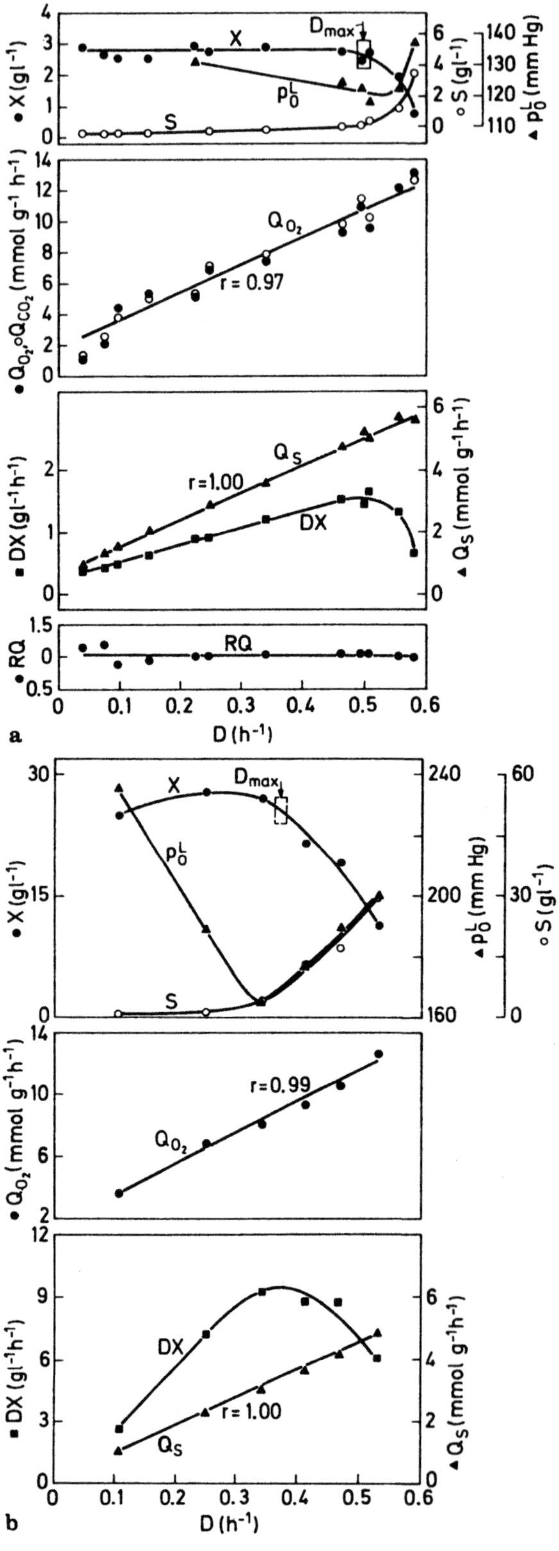
a
X
D_{max}
p_O^L
S
Q_{O_2}
r = 0.97
Q_S
r=1.00
DX
RQ
• X (g l⁻¹)
○ S (g l⁻¹)
▲ p_O^L (mm Hg)
• Q_{O_2}, ○ Q_{CO_2} (mmol g⁻¹ h⁻¹)
■ DX (g l⁻¹ h⁻¹)
▲ Q_S (mmol g⁻¹ h⁻¹)
• RQ
D (h⁻¹)
b
r=0.99
• Q_{O_2} (mmol g⁻¹ h⁻¹)

Fig. 17a and b

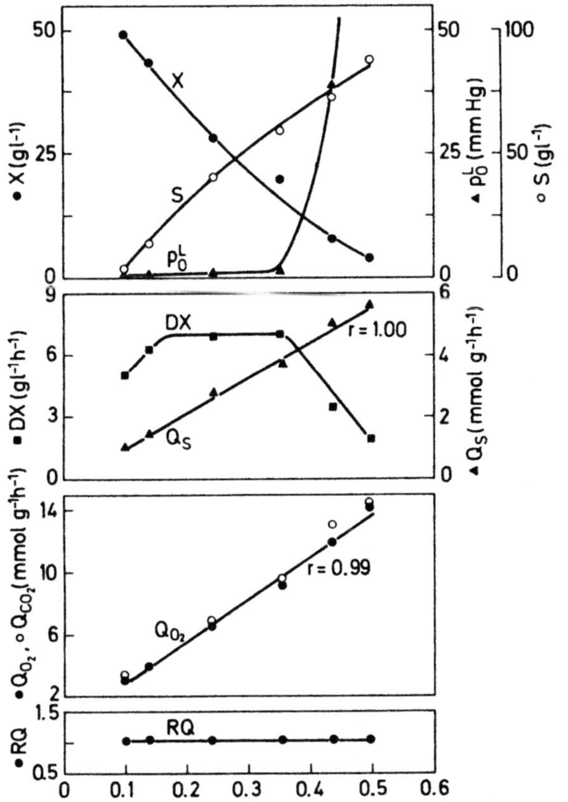

c

Fig. 17a–c. Effects of medium concentration in biotesting aeration/agitation devices. CSTR 7 l vessel with FBT stirrer. Growth of *Trichosporon cutaneum* in molasses medium of various strengths (Janshekar [36])

Fig.	S_0, g l^{-1}	vvm	rpm
17a	4.9	1	1200
17b	48.5	1	1600
17c	97.0	1	1600

r = coefficient of last square regression of Q_{O_2} and Q_s, respectively, against D. Gas for aeration was topped up with pure O_2 to prevent O_2 limitation. Increase of medium concentration shows decrease of D_{max} and relative decrease of productivity (DX). This is a typical picture generally interpreted as a result of limitation due to inadequate agitation/aeration. However, a more detailed analysis and improvement of medium composition showed that the more concentrated media used in experiments of Fig. 17b and c were inadequate in composition and preparation procedures

Unrestricted function of the entire set of active enzymes in a cell is therefore expressed by the Monod equation. Deviations indicate that the regulatory situation has become different from unrestricted growth or metabolic functioning. This is normally due to the presence of inhibitors(s), recalcitrant substrates or unknown limitations by nutrients.

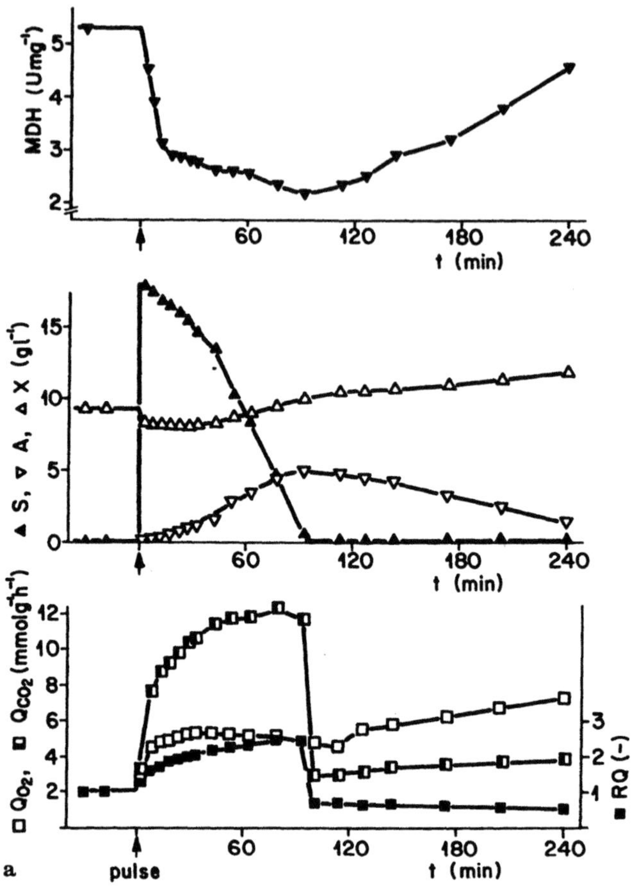

a

Fig. 18a. Regulatory pattern for malate dehydrogenase (MDH) activity in a pulse experiment with *Saccharomyces cerevisiae*. Glucose is pulsed to a continuous cultivation during steady state at $D = 0.10\ h^{-1}$ and $s_0 = 2\%$. Response of MDH-activity is quick indicating a mechanism acting on a allosteric level. Recovery from the pulse after exhaustion of the added glucose appears as a slow-process. Typical phenomena are the immediate increase of CO_2-production after glucose addition and the accumulation of ethanol during the next 100 min. It is significant that the specific oxygen uptake rate barely decreases under the condition of ethanol production. This indicates that true carbon limitation in the medium is maintained during the entire transient state after pulsing (Brändli [40])

Competitive and noncompetitive inhibition can be easily identified in the respective plots of chemostat experiments (Pirt [35]). The occurrence of this type of regulatory pattern, however, needs assessment of the presence of the actual inhibitor. Identification of the inhibitor is necessary for proper classification of the situation.

It must be mentioned that high values for ks are always suspect because they may indicate incomplete medium composition (hidden limitations) rather than the presence of an inhibitor. Similarly, recalcitrant substrates may exert effects on the modeling of growth kinetics similar to assumptions made for the presence of inhibitors. Finally, the inability of the cells to proliferate or form products caused by genetically encoded factors may lead to high values of K_s app.

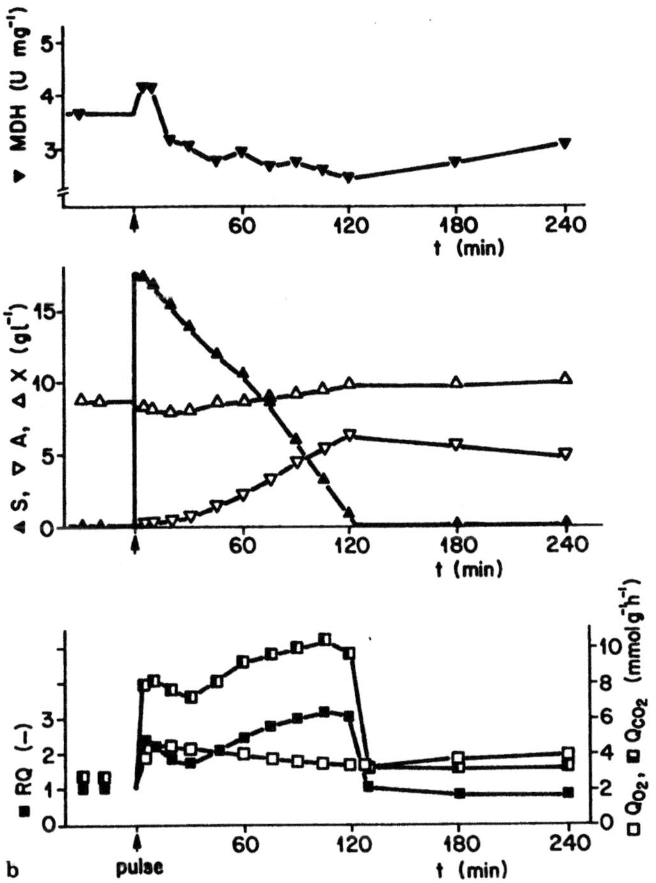

Fig. 18b. Regulatory pattern for malate dehydrogenase (MDH) activity in a pulse experiment with *Schizosaccharomyces pombe* in a defined C-limited medium (Table 13, D_w). Glucose pulse is done according to the experiment of Fig. 18a. A typical peak of MDH-activity is followed by repression during several hours indicating complex short- and long-term regulatory mechanisms for the enzyme. Again, specific oxygen uptake is essentially constant after the glucose pulse due to maintenance of carbon limitation under all conditions of the experiment. Decrease of Q_{O_2} in this yeast in the presence of excessive glucose is more a result of iron limitation than of "glucose repression" as demonstrated by the author (see also Fig. 11) (Brändli [40])

Besides growth kinetics, mass balancing of a bioprocess is also necessary. It provides the final testing methods for product formation and for all kinds of changes in the metabolic performance of the cell. A proper mass balance not only informs on the distribution of products but also gives proof of the accuracy of analytical work done on-line and off-line. Subsequent testing of the material balance during growth is of utmost importance in kinetic studies of microbial behavior.

Even though data from batch runs may be informative, in practice, they are less suitable because of reduced reliability, accuracy, and reproducibility of data. The accurate interpretation of simple batch experiments is much more difficult and critical. As a result, studies for the evaluation of the kinetic potential of the metabolism are prerequisite. This has been recognized for quite some time and was first applied for strain selection studies with chemostat methods [116].

Studies on growth kinetics using chemostat methods are most effective for the elucidation of influence of all kinds of extracellular parameters [117]. In the following two examples, shown in Fig. 17a–c and Fig. 18, interference caused by inadequate medium demonstrates the validity of this statement.

Figure 17 reports an experiment in which the medium concentration was increased sequentially. The concept of varying medium strength is the simplest approach in bioreactor testing because visualization of occurring mass transfer limitations can be best achieved in concentrated nutrient solutions. Upon reaching the critical potential for agitation/aeration, e.g., mass transfer, values for yield, respiratory quotient and kinetics become altered, whereas μ_{max} should remain constant.

This is therefore what one would expect to see in Fig. 17. However, the observed changes in this direction are not quite clear. When s_0 is increased in three independent experiments from 3% to 5% and finally to 10%, the behavior of the yeast strain *Trichosporon cutaneum* according to the specific oxygen uptake rate qO_2 is exactly as one would expect, e.g., a strong relationship to specific growth rate μ and not at all to s_0, although yields decreased drastically with increasing medium concentrations (and increasing D). Typically, the shapes of the plots for x and s become reduced and elevated, respectively, at higher dilution rates. Altogether, the result is increasing values for K_s app. In most cases, high apparent K_s values stand for a hidden limitation caused by inadequate dosing of medium supplements. In his work, Janshekar [36] was unable to identify the actual limitation factor in molasses media. However, in later experiments, he and other authors (Kaetterer [37]), carried out respective studies, designing a defined carbon-limited medium. This has a different dosage of several nutrients and is now used as a test medium for bioreactor testing. Mass transfer limitations for *Trichosporon cutaneum* in this medium induce a pure reduction of μ and no other changes such as yield or μ_{max} [38].

The second example of the importance of kinetic studies in metabolic and enzymatic investigations is represented in Fig. 18. It shows the activity of malate dehydrogenase in function of dilution rate as well as the usual plots of x, s, q_{O_2}, q_{CO_2} and RQ of the yeast *Schizosaccharomyces pombe* [39,40]. This system ought to work for the preparation of the two isoenzymatic forms of malate dehydrogenase. Even in extensive studies using a modified medium of Wickerham, stability of the chemostat states was always poor. The cell material harvested from this preparation appeared to be inconsistent, carrying varying fractions of the enzyme from one preparation to the other. Also, banding of fractions on the gels and in separation columns was barely controllable, showing that metabolic control of enzyme synthesis is very inconsistent.

A thorough study of the growth system e.g., strain and medium ended in the demonstration of a strong deficiency of trace elements, especially of iron. Fig. 19 shows the very different performance of the strain in revised medium WB, which allowed the perfected study of the enzymatic regulation of malate dehydrogenase.

These two examples show that more sophisticated work on medium design is required. Purely empirical approaches in studies of metabolic control, enzymes or process development are no longer suitable. More and more efforts are being undertaken to improve the composition of media and to adjust to the needs of the cells. Kinetic studies can be very helpful in this respect. This has been demonstrated among

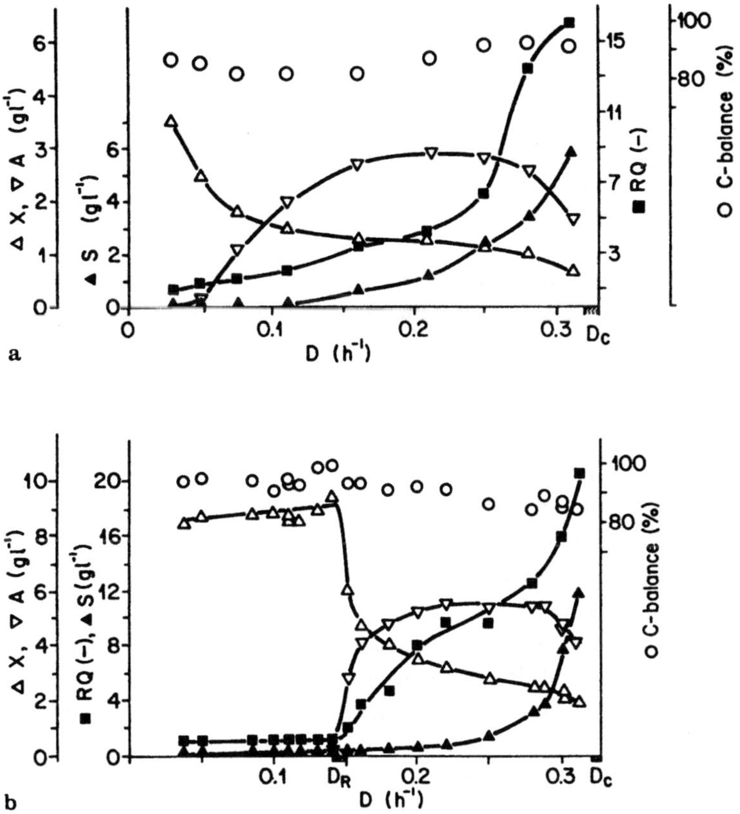

Fig. 19a and b. Chemostat experiments are excellent tools for testing the effects of chemical parameters. Biomass formation and residual glucose concentration plus respiratory quotient and mass (carbon) balance in function of the dilution rate are most informative for medium design. The fast rising values for s at $D > 0.1$ (h^{-1}), the low biomass formation at low D and the unspecific RQ data lead to the conclusion that the medium was not carbon limited (Table 13). Systematic medium design for this yeast yielded a regulatory pattern similar to that of *Sacch. cerevisiae* (see Fig. 18a). The composition of the medium D_w (Table 13) used in these experiments shows that primarily iron but also other trace elements are limiting in the modified Wickerham medium [40]. The comparison of the two sets of data from the same organisms shows the advantages of proper medium design for regulatory studies. Interpretation of Fig. 19a will create considerable difficulties and the cell material from such cultures is unsuitable for biochemical investigations

others by Dostalek et al. [41], Haeggstroem [42], Mateles and Battat [43], Kuhn et al. [44], and Goldberg and Er-el [45]. Fine overviews of growth kinetics, yields, energy balances and mass balancing for various substrates have been published by Mateles [46] and Abbott and Clamen [47].

A concept for using systematic approaches in medium design will be given later in the text. Thus, time-consuming experiments and trial and error procedures are no longer necessary and the endless testing of possible supplement combinations in shake flasks can be replaced by more efficient methods.

3.2 Composition of Media

3.2.1 Principles

Media composition plays a fundamental role in every growth experiment and bioprocess. The availability of necessary nutrients are decisive for the performance of the cell on both the level of metabolism and metabolism control. In other words, mass balance and bioreaction kinetics are governed by the chemical nature of the nutrient solution.

It is textbook knowledge that all metabolic functions are bound to the presence of water. Therefore, the solubility of the reactants necessary for growth has physiological and practical consequences.

Low solubility coefficients can be responsible for growth limitations, as well as for the formation of insoluble complexes due to the occurrence of chemical reactions during medium preparation. Additional problems are encountered if supplements have certain characteristics such as a strong lipophily, as in the case of hydrocarbons, or if solid materials are used as the carbon and energy source.

The objective of a medium composition study is twofold:

— to identify all components that are necessary for balanced growth or that may interfere with growth kinetics in some way: *qualitative aspect.*
— to establish the concentration of the components in an appropriate manner for the definition of the limiting substrate: *quantitative aspect.*

When a nutrient solution is chemically definable, it is called a synthetic medium. Because there are such a large number of medium components, they have been classified into the following four categories.

1. Substrate. This usually refers to the component which is added in the greatest quantity and which represents the limiting factor in the growth system. It usually consists of a carbon source, but may also be oxygen, nitrogen, or phosphorus, and may be considered as growth substrate. Media are often designed to be C-limited and mostly glucose represents the carbon and energy source.
2. Nutrients (or macro elements). These are components that are used in large amounts, such as C, O, H, N, S, P, K, Ca, Mg. These elements are supplied in a form that may be organic (carbon source or nitrogen source) or inorganic. Selection of the proper form must be carefully studied for biological and practical reasons (solubility).
3. Trace elements (or micro elements). These are mostly metals, such as Fe, Mn, Mo, Zn, Co, Ni, V, B, Cl, Na, Si). These elements are usually supplied in the form of simple salts that are dissolved in minor amounts of milligrams or even micrograms per liter. However, when trace elements are dissolved in water and then heated, new salts that are insoluble may easily form. Although the essential atoms would be present in the final solution, they would not be accessible, thus giving rise to faulty results. Therefore, it is necessary to follow procedures carefully in media preparation.
4. Growth factors. These are mostly vitamins — organic compounds that are supplied in low quantities but cannot be synthesized by the cell; in other cases, growth factors may limit growth kinetics. In medium preparation, the problem with organic growth factors is their susceptibility to degradation during heating.

Consequently, such growth factors must be dissolved and (filter-)sterilized separately and then added only after the bulk of the media has been prepared.

There is a great deal of literature available on the vitamin requirements of bacteria and yeasts. Much of it has been summarized by S. A. Koser [48].

Amino acids, or building blocks for protein synthesis, must be added to media when auxotrophic strains are used or when the potential of amino acid synthesis is restricted. The latter case creates problems in proper identification of the actual bottleneck due to the complicated regulation patterns in these branched, biosynthetic pathways.

In many cases, media development does not meet the final requirements of a synthetic solution because complete identification of all essential compounds is not possible. In such cases, complex media that make use of yeast extract, yeast lysate, peptone, or meat extract, etc. have been developed. These supplements have the advantage of providing 'good' growth, although they render a more difficult situation in the elucidation of kinetic, regulatory, analytical and preparative work.

At the beginning of every systematic medium development procedure is the identification of the nutritional demands of the strain. In order to determine the essential molecules and ions required for full growth, the composition of the cell itself must first be evaluated. By using available analytical data on the composition of the cell, a list of its essential atoms and their concentrations can be produced. This is then followed by the selection of the suitable form of salts and the calculation of their amounts. Examples of some media ingredients are given in the following tables.

The composition of the biomass formed is the basis for all mass considerations in growth experiments. Similarly, the ratios of the major atoms such as C, H, N and O must be met by the medium. Table 5 gives a typical example of an overall analysis of baker's yeast and fodder yeast which shows minor differences between the two strains obtained from different media. A more recent publication gives typical data in more detail, including categories of trace elements and vitamins [114, 127].

Typical values for the ratios of inorganic nutrients for carbon-limited growth of yeasts and many bacteria are given in Table 6. This type of listing is most helpful for the evaluation of media or in experiments for designing defined media.

The following Table 7 (for yeast) and 8 (for *Escherichia coli*) show a synopsis of supplement considerations made from a random sampling of data from the literature of the last few years. These tables show the great variations in the form and amounts recommended for supplements of basal media. The impression results that in many cases the medium composition may be far from optimal, inducing more or less unbalanced cell growth.

Table 6. Supplements of nutrients per 100 g glucose for carbon-limited yeast growth. This type of listing is most helpful for evaluation of media

Atom	Amount (g)	In form of
N	5.6	$(NH_4)_2SO_4$ and $(NH_4)_2HPO_4$
P	1.5	$(NH_4)_2HPO_4$
K	1.52	KCl
Mg	0.148	$MgSO_4 \cdot 7\,H_2O$

Table 7. Synopsis of media used for yeast growth experiments. Arbitrary selection of 8 papers reporting *Candida lipolytica*, *Candida tropicalis* and two strains of bacteria used for hydrocarbon media [91–98]

Components	Form	Amount per liter	Reported times
Ca	$CaCl_2$	50 µg—500 mg	4
Cu	$CuSO_4$	80 µg	3
Fe	$FeCl_3$, $FeCl_4$, $FeSO_4$	1 µg–70 mg	5
K	KCl, KI, K_2HPO_4, KH_2PO_4	200 µg–7 g	9
Mg	$MgCl_2$, $MgSO_4$	0.2 g–10 g	8
Mn	$MnCl_2$, $MnSO_4$	30 µg–100 mg	7
Na	NaCl	50–100 mg	2
Mo	Na_2MoO_4	10 µg	1
NH_4	$(NH_4)Cl$, $(NH_4)_2SO_4$, $(NH_4)_2HPO_4$	0.2 g–8 g	10
Zn	$ZnSO_4$	80 µg–6.99 mg	3
B	H_3BO_3	200 µg	1
Biotin		23 µg	1
Ca-pantothenate		23.3 mg	1
m-Inositol		46.6 mg	1
Phenol-red		1 mg	1
Pyridoxin		1.16 mg	1
Thiamine		3 µg–4.66 mg	2
Yeast extract		100 mg	2
Substrate	n-dodecane, n-tetradecane n-hexadecane, gas oil	10–15 g	8

Table 8. Synopsis of media used in experimentation including mass culture of *Escherichia coli*: Included are 15 reports [99–112] from the years 1951–1980. The list indicates the arbitrariness in the selection of elements, their forms, and concentrations of the carbon source. It also shows some attempts for supplementing some organic factors. Obviously arbitrary mixtures cannot maintain C-limited growth

Component	Form	Amount per liter	Reported times
Carbon source	Glucose, glycerin casein	2–33 g	15
Amino acids	Casein	6.7–30 g	2
Growth factors	YEPD	0.5–31 g	5
K	K_2HPO_4, KH_2PO_4	1 g–13.6 g	22
Na	$NaHPO_4$, NaCl, Na_2SO_4	11 mg–7 g	13
$(NH_4)_2$	$(NH_4)_2HPO_4$, $(NH_4)_2SO_4$, NH_4Cl,		14
	NH_4NO_3	0.8 g–8.0	14
Mg	$MgCl_2$, $MgSO_4$	2.4 mg–2 g	15
Na	NaCl, $NaSO_4$	11 mg–3 g	9
Ca	$CaCl_2$	1 mg–40 mg	8
Fe	$FeSO_4$, $FeCl_3$	0.25 mg–40 mg	8
Cl	NH_4Cl, $FeCl_3$, $CaCl_3$, $NaCl_3$, $MgCl_3$	0.5 mg–3 g	21
Trace elements			
Antifoam			

A more difficult problem in the given approach for developing media composition, however, is finding data on the nature of molecules present only in extremely small quantities. Trace elements may play an important role in metabolism, but they cannot always be easily estimated analytically. This is one of the causes of non-Monod growth behavior. Instead of spending much time and effort trying to analytically determine all of these minor components, a more efficient method would be to do growth experiments using pulse techniques in chemostat studies.

3.2.2 Examples of Media Ingredients

Many recipes for media have been developed and described in the literature. However, in only a few cases have the limiting factors been clearly defined. To create a media, one need only examine examples in the literature and then add or omit components according to the maximum growth rates possible or obtained.

There is a large amount of data available in the literature for the development of media. For yeasts, the review article by Suomalainen and Oura [49] on essential nutrients, trace elements and growth factors should be consulted. This review also lists the factors required for anaerobic growth and, most important, natural and industrial sources of yeast nutrients such as molasses, dough, malt worts, grape must, sulphite waste liquor, and hydrocarbons. Another example for extensive nutrient lists for general culture work has been published in a book edited by Praeve et al. [50]. Using tables of compiled data from these sources, medium for growth can be developed easily. However, for more reliable work, a detailed study of the growth requirements of the cell will be necessary and a thorough testing for built-in limitations must be undertaken.

A final example of this kind of a media for cellulolytic fungi is given in Table 9. Many recipes found in the literature appear suitable for the growth of this kind of organism. However, systematic testing has shown that none of these media can provide full yield at a fast growth rate, and with some strains, intermediate metabolites such as acetic acid may form due to artifactual regulation.

Table 9. Proposed composition for media used in cultivations for cellulolytic fungi. Concentration in g atoms for 1 % cellulose media

Organism	P $g\,l^{-1}$	N $g\,l^{-1}$	S $g\,l^{-1}$	K $g\,l^{-1}$	Mg $mg\,l^{-1}$	Ca $mg\,l^{-1}$	Fe $mg\,l^{-1}$	Ref.
Trichoderma reesei	0.45	0.45	0.38	0.57	29	82	1.0	119)
Sporotrichum pulverulentum (= *Phanerochaete chrysosporium*)	0.2	0.4–1.0	0.07	0.26	49	20	2.7	120)
Phanerochaete chrysosporium	0.23	0.34	0.07	0.59	49	27	2.0	121)
Trichoderma reesei	0.27	0.85	1.0	0.34	49	—	2.0	122)
Polyporus versicolor	0.60	0.88	0.07	0.76	49	4	trace	123)
Chaetomium cellulolyticum	0.23	0.33	0.07	0.55	49	—	2.0	124)

3.2.3 Preparation of Media

As has been indicated, the dissolution and mixing of supplements may create problems of precipitation. Heat sterilization may also lead to an unsuitable preparation. In any case, these purely practical aspects are of great significance, and it is likely that erroneous results have resulted in some experiments due to superficial treatment of this aspect.

The formation of insoluble salts and their subsequent precipitation is particularly important in cases of trace elements where the sediment is only barely visible. Iron, which is an important component of all haeme-type cytochromes because of its effect on respiratory activity, is such a component that is easily lost (see, for example, Sect. 3.4.).

This aspect of losing trace elements, of course, has been known for a long time. In an excellent review published in *Inorganic Nutrition*, Hutner [51] states that not only are the requirements for trace elements too often overlooked, but also that the mode of media preparation is inadequate (see Table 10). Hutner notes that chelate formers are necessary to prevent the formation of insoluble salts, mostly in connection with phosphate or magnesium. Citrate is a good example of an iron-binding component — a less effective chelating agent is histidine and other amino acids. The suitable selection of salts is also important — in some cases, inorganic phosphates may be replaced by organic forms such as glucose phosphate or nucleotide phosphate. For the proper elucidation of trace element requirements, Hutner recommends the preparation of a mixture of trace elements (according to Table 9) that is titrated against a constant chelate former.

Another problem in media preparation is the degradation of heat-susceptible molecules. They usually must be dissolved in small volumes of water, sterilized by filtration and added separately to the final medium.

Yet another aspect of practical medium preparation is the solubility of the gas in the microbial culture medium. This solubility depends not only on the nature of the gas, but also on the concentration of salts present in the mixture. Oxygen, in particular,

Table 10. Trace element mixture according to Hutner [51]

Element	Final medium mg l^{-1}	Salt	Gravimetric	
			Factor	Salt g
Fe	6.0	$Fe(NH_4)_2(SO_4)_2 \cdot 6\ H_2O$	7.0	42.0
Mn	5.0	$MnSO_4 \cdot H_2O$	3.0	15.0
Zn	5.0	$ZnSO_4 \cdot 7\ H_2O$	4.4	22.0
Mo	2.0	$(NH_4)_6MoO_7O_{24} \cdot 4\ H_2O$	1.8	3.6
Cu	0.4	$CuSO_4$ (anhydr.)	2.5	1.0
V	0.2	NH_4VO_3	2.3	0.46
Co	0.1	$CoSO_4 \cdot 7\ H_2O$	4.8	0.48
B	0.1	H_3BO_3	5.7	0.57
Ni	0.1	$NiSO_4 \cdot 6\ H_2O$	4.5	0.45
Cr	0.1	$CrK(SO_4)_2 \cdot 12\ H_2O$	9.6	0.96
				87.02

has poor solubility, even in salt-free water, so that the chances are high that this 'nutrient' will become limiting in more concentrated media. The Deckwer Group (Schumpe et al. [18]) has studied this problem of gas solubility with the following results:

In media with high salts concentrations, problems of solubility or oxygen supplementation may become critical. The higher the salt concentration, the greater the probability that insolubility will occur (see Figs. 17a–c, Janshekar [36]). In these figures, a drastic modification of Monod kinetics in the growth of *Tr. cutaneum* takes place. The best explanation for this is that the difference in glucose concentrations as 5% and 10% runs in otherwise identical experiments accounted for incomplete yield at increasing growth rates.

The form of the carbon source may create some problems, as for example: solids (starch), immiscible liquids (hydrocarbons or toxins), methanol, or ethanol. In these cases, separate additions under correctly chosen conditions may be the solution to the problem.

It is not possible to make a general recommendation on the best procedure for media preparation. Every process and cultivation has its own peculiarities and appropriate layout for medium supplements and sterilization. Fortunately, improvements in filtration techniques has been of great advantage, especially as a replacement for classical heating methods in autoclaves.

3.3 Systematics of Media Design

The correct designing of a medium that provides full cell synthesis capacity needs improvement. The usual approach for the selection of essential components and their correct amounts by using shake flask experiments based on trial and error is very empirical, time-consuming, and unreliable. This also holds true for simple batch cultures in agitated vessels, which very often lack the necessary accuracy and reproducibility for proper interpretation.

Chemostat methods are more efficient, as has been shown by Mateles and Battat [43]. These authors were probably the first to make use of this efficient method for the development of carbon-limited media. A useful review on improvements in the methodology of medium design has been published by Goldberg and Er-el [45]. They also emphasized the usefulness of chemostat experiments and cited the pulsing method developed by others.

The pulsing of a potential nutrient, trace element or growth factor into a culture during the steady state leads to an immediate response as shown in Fig. 20a. Such a response can be very informative. Nonspecific reactions to an addition indicate the ineffectiveness of the nutrient, whereas the exponential increase of biomass after a pulse gives proof of a hitherto unknown limitation. As shown in the Fig. 20a, pulsing can be elaborated in order to assess the necessary amounts of supplements. This type of approach is most efficient and even rather complex problems may be solved in a short amount of time. For example, amino acid limitations that by any other method may be difficult to determine can be unraveled in a systematic way. Using this technique, amino acid auxotrophy or low biosynthetic capacity can be easily detected. A perfect example of the detection of limiting amino acid synthesis in a

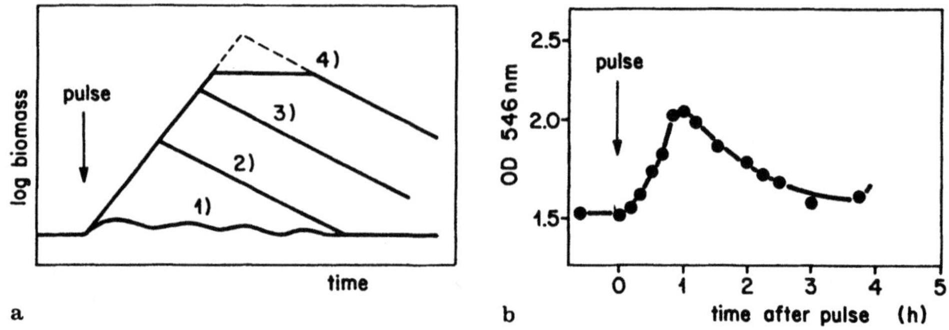

Fig. 20a and b. Pulsing of potential media components to a chemostat culture during steady state is a most efficient method for media design. **a.** Time course of biomass after pulse: 1) "unspecific" chances indicate that the substance in question had no nutritional effect; 2), 3) true effects on growth exert exponential kinetics; 4) excessive dosage; **b.** Actual experiment with *Bacillus caldotenax*. Response of optical density OD to a pulse of methionin into an incomplete defined medium. Demonstration of an auxotrophy. (Kuhn et al. [44])

strain of *Bacillus caldotenax* with chemostat pulses was demonstrated by Kuhn [44]. The auxotrophic nature of this strain became clear after a pulse with casamino acids. In a second round of pulses, combinations of four to five amino acids were tested with the result that the groups containing methionine and glycine had positive results. Finally, members of these two groups were tested in a third round of pulses, with the result that methionine addition to the final medium (see Fig. 20b) turned out to be essential and glycine caused improvement in the yields.

A concept for systematic medium design is consequently composed of the following six steps:

STEP 1: Selection of essential components

The main essential supplements for a medium are known mostly from preliminary work. Also, valuable information is available from the literature, e.g., for bacteria (Schuegerl [5]), for yeasts (Suomalainen and Oura [49]). Analytical data on dry matter, as discussed earlier, are also of great help for designing an initial usable composition. One should note that besides nutrient salts, growth factors, and trace elements, the gas phase is also of importance (e.g., CO_2). In essence, a defined medium is composed of a carbon and energy source, nutrients, trace elements and growth factors.

STEP 2: Preparation of the medium

Normally, the necessary medium components are dissolved in water and mixed to make up the final growth solution. This method is acceptable as long as no chemical reactions lead to the formation of precipitates and no heat susceptible substances must be added. Otherwise, critical components must be sterilized separately and added later.

Usually in this phase of medium preparation, very careful consideration of the chemical aspects must be made. This is in view of possible studies in more concentrated media. It is most important to check the final medium for possible precipitations.

STEP 3: Preliminary x-D-diagram

The plots of biomass (x) and limiting substrate (s) from a chemostat experiment are most informative (Y, k_s, μ_{max}). They show whether or not the growth kinetics fit the Monod-model. Also, turnover rates of substrate and rates of oxygen uptake, carbon dioxide production and the respiratory quotient should be calculated, plus the necessary mass balance. Evaluation of the data indicates in most cases whether or not the medium provides full growth.

STEP 4: Assessment of carbon limitation

The correct concentration of the carbon source to provide a carbon-limited medium is found according to the usual method described by Pirt [35] by variation of the ratio of C-source to the mixture of nutrients, trace elements and growth factors.

STEP 5: Optimization

The carbon-limited medium is tested by chemostat pulse/shift experiments for possible hidden limitation that were not visible in the batch experiments of Step 4.

STEP 6: Final x-D diagram

The range of dilution rates permitted is found with the final composition of the defined medium. A complete diagram for x, s, q_p and other metabolic products formed is constructed in function of D. Then q_s, q_{O_2}, q_{CO_2}, q_p, and RQ are calculated and the values obtained for Y, K_s, and μ_{max} are compared with the information known about the strain. The mass balance together with the other data provides valuable insight into the regulatory pattern during growth.

The given procedure is not only prerequisite for consistent regulatory investigations, but also for the proper identification of the parameters influencing microbial growth.

3.4 Examples of Carbon-limited Media

A perfect medium composition is prerequisite for any biological study. Examples of the results of complete medium design will be given in this section and illustrated by some biological and engineering data.

Yeasts have been the object of many scientific investigations, and there is a considerable amount of data on mutable media compositions. However, balanced growth using such classical media could barely be attained. Instability of the steady state in continuous cultivations, poor reliability of enzymatic data, and difficulties in purification work of isoenzymes were the consequences of this situation.

An example of perfect carbon limitation under all circumstances of yeast growth is given in Table 11. This medium is composed of chemically defined substances, except for a very minor supplement of yeast extract. This compromise indicates the

Table 11. Example of a complete synthetic carbon-limited yeast medium containing nutrients, growth factors and trace elements for full growth and a small amount of complex components such as yeast extracts (e.g. see [53])

Supplements per 10 g glucose:	
2 g $(NH_4)_2SO_4$	3.5 mg $MnSO_4 \cdot 2\,H_2O$
0.64 g $(NH_4)_2HPO_4$	0.01 mg Biotin
0.29 g KCl	20 mg m-Inositol
0.15 g $MgSO_4 \cdot 7\,H_2O$	10 mg Ca-Pantothenate
0.14 g $CaCl_2 \cdot 2\,H_2O$	2 mg Thiamin
0.78 g $CuSO_4 \cdot 5\,H_2O$	0.5 mg Pyridoxin
4.8 mg $FeCl_3 \cdot 6\,H_2O$	
3 mg $ZNSO_4 \cdot 7\,H_2O$	Yeast extract 0.25 g

still unidentifiable auxotrophy present in yeast strains even when they are normally considered as prototrophic.

Very often, defined media for yeast studies are actually controlled by ân unknown limitation of trace elements such as iron, magnesium or manganese that override the effects of carbon limitation. The results of this inconsistency can be rather drastic, as demonstrated by Rieger [53]. When an improved medium with perfect C-limitation was used in regulatory studies of *Saccharomyces cerevisiae*, differences between earlier experiments became obvious. The classical theory of control of respiration by glucose turned out to be erroneous. It was shown that the rate of respiration is linearly bound to the growth rate, making a maximum of approximately $\mu_{max}/2$. The q_{O_2max} remains constant even under conditions of 'glucose repression' and is not, as believed earlier, reduced as 'repression' increases (Fig. 21). The consequences of this faulty medium composition are very far-reaching and will be described elsewhere. The practical consequences of these finding, however, are obvious. Trace elements are good candidates for producing hidden limitations and the respiratory sequences of metabolism are the primary targets of their controlling effects.

Another example of erroneous results caused by hidden limitations of trace elements (iron) is shown in Fig. 19. *Schizosaccharomyces pombe* is believed to be a special case for metabolic control by glucose. As shown in Fig. 19a, yield of this fission yeast was thought to be very low at all growth rates in continuous cultures. However, interpretation of such a growth pattern was quite difficult because the repression became apparent at all growth rates down to $D < 0.1\ h^{-1}$. True release from repression was possible only at very low growth rates. However, reproducibility of the steady states was poor, as was stability of biomass concentrations. Revision of the medium yielding true carbon limitation changed the performance of *Schizosaccharomyces pombe* entirely to the well-known regulatory pattern of *Saccharomyces cerevisiae* (Fig. 19b). There is now a definable maximum flux of substrate for oxidative turnover and glucose supression beyond this point. This more correct regulatory pattern has not yet been characterized in every respect. However, the plots of malate dehydrogenase found in these cells also show that there are different patterns of activity in the two media. These latter data demonstrate perfectly how much enzymatic investigations depend on the consistency of extracellular parameters of growth.

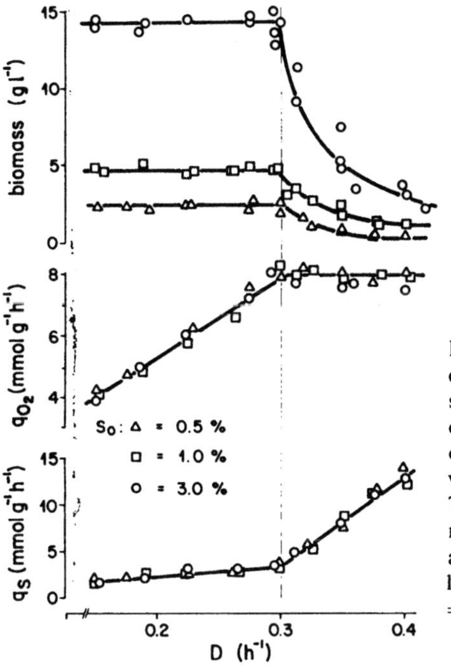

Fig. 21. Test for adequate mixing, aeration and true carbon limitation. The final proof for the proper solution of the extracellular conditions is the use of continuous cultivation at various concentrations of the medium. The regulatory pattern must invariably remain the same as it is given in this figure. The strain of *Saccharomyces cerevisiae* has a critical rate of glucose uptake for onset of glucose repression at q_s = 3.5 mmoles g^{-1} h^{-1} = constant for s_0 = 0.5, 1, 3%. This critical rate was found at D_R = 0.30 h^{-1} = constant. For details see Rieger [53]

Generally, *bacteria* have a much wider media requirement spectrum than yeasts, and only a few correctly-developed media for mass culture exist. Table 12 gives an example of a defined C-limited medium for *Escherichia coli*. This medium was initially developed for interferon production in genetically engineered strains. In *E. coli* interferon production was strongly bound to growth and thus depended primarily on fast balanced growth behaviour [52].

Figure 22 shows chemostat data obtained with a wild B-type strain of *E. coli* and a K 12-strain genetically engineered for interferon production. It is interesting to note that the acetic acid formation observed in these strains seems to reflect a constitutive situation. Later experiments have shown, however, that as in the case with yeasts, restrictions of trace element delivery to the cell were responsible for the secretion of C_2-compound. Again, an example for the previous statement is that trace elements — particularly Fe — are most critical in correct feeding of the cell and that the target of such inconsistencies are primarily the respiratory sequences, including the final steps catalyzed by cytochromes.

The medium composition given in Table 13 is the result of investigations of *Schizosaccharomyces pombe* growing on glucose. This system has been thoroughly tested for obligate medium supplements trace element limitation [40]. It is known from this work that faulty information was previously obtained because the added iron was physiologically lost by the precipitation of insoluble salts formed during medium preparation. The selection of appropriate pH in the storage vessel and separate addition of critical salt fractions may improve the obtainable respiratory capacity and maintain maximum q_{O_2}-values even in presence of glucose.

Table 12. Composition of media for mass culture of *Escherichia coli*. The media are designed for true carbon limitation applying pulse method in continuous cultivation. See Fig. 22a, B for diagrams obtained with a wild type strain B/r and K12 D1 in the complex medium. For details see [52]

Defined for *E. coli* B/r

Component	$g\,l^{-1}$
Glucose monohydrate	4.4
NH_4Cl	3.22
KH_2PO_4	0.525
$MgSO_4 \cdot 7\,H_2O$	69.7×10^{-3}
K_2SO_4	21×10^{-3}
$FeSO_4 \cdot 7\,H_2O$	2.74×10^{-3}
$CaCl_2$	0.8×10^{-3}
Trace elements*	0.5 ml
Antifoam (PPG)	0.5 ml

Complex for *E. coli* B/r

Component	$g\,l^{-1}$
Glucose monohydrate	4.4
Casamino acids	5
Yeast extract	1
NH_4Cl	3.22
KH_2PO_4	1.05
$MgSO_4 \cdot 7\,H_2O$	136×10^{-3}
K_2SO_4	42×10^{-3}
$FeSO_4 \cdot 7\,H_2O$	3×10^{-3}
$CaCl_2$	1.6×10^{-3}
Trace elements*	1 ml
Antifoam (PPG)	0.5 ml

$AlCl_3 \cdot 6\,H_2O$	20
$CoCl_2 \cdot 6\,H_2O$	8
$KCr(SO_4)_2 \cdot 12\,H_2O$	2
$CuCl_2 \cdot 2\,H_2O$	2
H_3BO_4	1
KJ	20
$MnSO_4 \cdot H_2O$	20
$NiSO_4 \cdot 6\,H_2O$	0.9
$Na_2MoO_4 \cdot 2\,H_2O$	4
$ZnSO_4 \cdot 7\,H_2O$	4

Defined for *E. coli* K12 D1

Component	$g\,l^{-1}$
KH_2PO_4	0.89
NH_4Cl	6.6
$MgCl_2 \cdot 6\,H_2O$	0.21
K_2SO_4	0.294
$FeSO_4 \cdot 7\,H_2O$	5.6×10^{-3}
$CaCl_2 \cdot 2\,H_2O$	1.46×10^{-3}
Histidine	0.192
Proline	0.234
Tryptophan	0.102
Glucose monohydrate	17.6 g
Antifoam (PPG)	0.5 ml
Trace elements*	0.5 ml

Complex for *E. coli* K12 D1

Component	$g\,l^{-1}$
Glucose monohydrate	3.3
KH_2PO_4	0.45
$MgCl_2 \cdot 6\,H_2O$	0.21
K_2SO_4	0.296
Casamino acids	5
Yeast extract	1
$FeSO_4 \cdot 7\,H_2O$	5.6 mg
Antifoam (PPG)	0.5 ml
Trace elements*	0.5 ml

* Solution, $mg\,l^{-1}$

A final example of the importance of correct medium design is the application of defined biosystems for testing aeration and agitation hardware. The basic idea in such an approach is the fact that chemically defined media support biologically defined regulation of growth and product formation. Variation of medium concentrations (s_0) provides the possibility of correctly adjusting the biosystem to the mass transfer capacity of the bioreactor (Fig. 22). In such a manner, maximum mass transfer rates for oxygen or other substrates can be measured. In this way, the performance of aeration/agitation can be characterized on the basis of the actual properties of process solutions. Typical data such as t_{mix}, k_La, velocity of liquid, productivity and other data in function of P/V are obtained for a given device (Table 14). Such sets of data allow the hardware designer to draw the necessary conclusions for further reactor design and improvement.

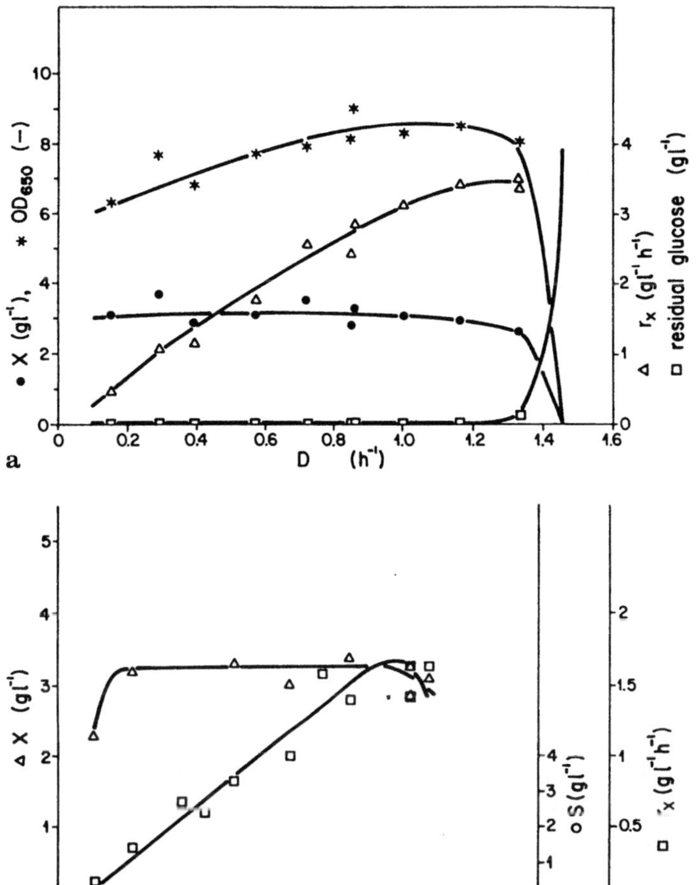

Fig. 22a and b. Medium design for mass culture of *Escherichia coli*. Descriptions of media for propagation of this bacterium are easily found in literature. However, most of them are poorly suitable for mass culture due to incomplete composition bearing hidden limitations. As a result poor yields of biomass, reduced values for μ_{max}, and formation of acetic acid at higher growth rates and apparently high K_s — values are obtained. A purely carbon-limited medium leads to typical data without any formation of metabolites on the C_2-level. **a.** *E. coli*, strain B/r, complex medium containing casamino acids plus some yeast extract (see Table 12). In the defined medium growth behaviour is similar to the values obtained with the complex medium. **b.** *E. coli*, strain K12, D1; Complex medium with glucose as a C-source (Table 12) [52]. Maximum specific growth rates are similar as with the defined medium of Table 12

Table 13. Development of a true carbon limited medium for *Schizosaccharomyces pombe* for aerobic growth (see [39]). The modified medium (first column) effected regulatory effects which could be poorly interpreted. Extensive study of the nutrient demand yielded in the composition given in the second column

Components	Modified Wickerham medium	Medium D_w	
$(NH_4)_2SO_4$	3.5 g l^{-1}		17.8 g l^{-1}
KH_2PO_4	1.0		5.7
$MgSO_4 \cdot 7\,H_2O$	0.5		1.2
NaCl	0.1		—
$CaCl_2 \cdot 6\,H_2O$	150 mg l^{-1}		150 mg l^{-1}
$ZnSO_4 \cdot 7\,H_2O$	0.4		30
$FeCl_3 \cdot 6\,H_2O$	0.2	$FeSO_4 \cdot 7\,H_2O$	100
$MnSO_4 \cdot 1\,H_2O$	0.4		32
$CuSO_4 \cdot 5\,H_2O$	0.04		0.79
H_3BO_4	0.5		15
KJ	0.1		1
$(NH_4)_6Mo_7O_{24} \cdot 4\,H_2O$	1.0	$Na_2MoO_4 \cdot 2\,H_2O$	5
$CoCl_2 \cdot 6\,H_2O$	—		5.6
+-Biotin	0.05 mg l^{-1}		0.1 mg l^{-1}
Ca-Pantothenate	1.0		20
Nicotinic acid	1.0		15
m-Inositol	10		160
Thiamine-HCl	1.0		4
Pyridoxin-HCl	1.0		10
L-Asparagin	1.5 g l^{-1}		—
Na-Citrat	—		0.5 g l^{-1}
EDTA	—		0.1 g l^{-1}
o-Phosphoric acid	—		0.75 ml l^{-1}
Glucose	10 g l^{-1}		40 g l^{-1}

4 Conclusions

This review deals with the possible effects of the environment on the behavior of the microbial cell during growth in submerged culture. The numerous effectors can be categorized into two classes of *physical* and *chemical* parameters.

The first category is strongly related to the containment in which growth takes place. Agitation and aeration must also meet the requirements for practical performance of the process, e.g., the exclusion of foreign cells (sterility), flotation, sedimentation of solids (cells), wall growth, loss of liquid (foam formation or excessive power consumption for mixing (in larger units). Some reactor configurations applied in biotechnology are described in Sect. 2. In their characterizations, emphasis is given to the present tendency to replace the classical STR and FBT with various types of loop reactors. Short loop configurations are especially capable of maintaining short mixing times. The appropriate layout, as found in the COLOR or TORUS types, also provides exact control of flow patterns in simple and more complex materials of newtonian and non/newtonian properties with strong tendencies for extreme foam formation. Mechanical agitation principles are best suited for an increase of power input, providing a good basis for the enhancement of productivity. An increase in the efficiency of all bioprocesses is the demand of the future, while

Table 14. Carbon-limited media suitable for bioreactor testing using *Trichosporon cutaneum*. Data for productivity r_x indicate the high potential for the increase of efficiency inherent to the compact loop forms like the COLOR and TORUS systems [38]

Reactor Configuration	$r_{x,max}$ $g\,l^{-1}\,h^{-1}$	OTR_{max} $g\,l^{-1}\,h^{-1}$	Oxygen supply from	Media	Aeration/ agitation	Refs.
STR 7 l	7	5.2	air	10% molasses	$n = 1200\ min^{-1}$ 1.0 vvm	[36]
STR 7 l	14	12.1	air + O_2	10% molasses	$n = 1600\ min^{-1}$ 1.0 vvm	[36]
STR 30 l	9.8	4.2	air	10% glucose	$n = 1000\ min^{-1}$ 0.9 vvm	[38]
COLOR 50 l[a]	20	14.5	air + O_2	14% glucose	—	[29]
COLOR 50 l	11.4	6.2	air	10% glucose	$n = 1000\ min^{-1}$ 1.4 vvm	[38]
COLOR 550 l	8.9	3.8	air	8% glucose	$n = 960\ min^{-1}$ 1.8 vvm	[38]
JLR 2300 l	5.3	2.1	air	8% glucose	1.2 vvm, pump	[38]
TORUS 100 l	7	4	air	7.5% glucose	$n = 1750\ min^{-1}$ 1.5 vvm	[20]
	19	10	air + O_2	7.5% glucose	$n = 1750\ min^{-1}$ 1.5 vvm	[20]

[a] *Candida tropicalis*

high process concentrations are prerequisite for better economy in downstream processing.

In most cases, the *chemical* effects of the cell environment are complex. Principles of medium development are given in Sect. 3, with examples of how limitations of nutrients act. Examples of some now recognized limitations — usually caused by the absence or loss of trace elements — are also given. Methods for the detection of such shortages are developed and described. Consequent use of chemostat methods (pulsing and shifting) are especially efficient tools for the proper identification of hidden limitations caused by the lack of essential components. This allows one to design media in a systematic manner, as it provides a sound basis for maintaining consistent metabolic control of the entire cell population.

The adequate selection of chemical parameters therefore is shown as a definite basis of behavior during growth. Thus, the composition of the media is responsible for the accumulation of these cells in a population that can develop optimally under the conditions selected. This concept — in view of the goals set — is of fundamental importance. The proper composition of media allows for the accumulation of desired mutants, as they are produced by nature, in a continuous culture. Nature can potentially provide an enormous number and variety of mutants, which is barely exploited, for example, in the area of environmental protection. Potent cell populations for thermophilic sludge treatment or recalcitrant compound degradation could be enriched out of a natural microflora. In addition, such enrichment concepts could simultaneously provide the necessary basis for process development and scale up, through the application of systematically selected parameters. A similar situation is encountered in populations from industrial production plants, in which extensive testing of the solution of media components would in many cases improve the maintenance of high productivities.

Finally, considerations of this kind are important in any case of the microbial mass culture of genetically engineered cells carrying genes of animal origin. Such constructions are normally equipped with an antibiotic resistance in the genome. The supply of a respective drug to the cultivation media would permit growth of cells with the plasmid bearing coupled genes for resistance and for product formation. Consequently, the growing population would consist exclusively of antibiotic-resistant and therefore producer cells. This is in accordance with the experience made in interferon production research. Needless to say, such a concept provides many possibilities for the development of reproducible, stable, and economical processes in both small and large scales.

Present knowledge of hardware configurations for the mass culture of microbes, and the effects of nutrients, growth factors, and trace elements on metabolism will increase as more attention is devoted to genetic engineering. Only basic knowledge of all of the components interacting with the growing cell can provide the prerequisites for scientific and application-oriented programs.

5 Acknowledgements

The assistance of Dr. Ingo Adler, Dr. Hans-Peter Meyer, Dr. Bernhard Sonnleitner, and Lee Anne Clark in the preparation of manuscript is appreciated. Funding for the major part of the underlying experiments was provided by the *Kommission zur Foerderung der Wissenschaftlichen Forschung*, Bern.

6 Symbols and Explanation

C-limitation	carbon limitation
C-limited	carbon limited
D	dilution rate (h^{-1})
D_c	critical dilution rate (h^{-1})
DDC	direct digital control
D_R	critical dilution rate for pure oxydative turnover (h^{-1})
k_La	mass transfer coefficient (liquid film) (kmol m^{-3} s^{-1})
K_s	saturation constant (mg l^{-1})
N	revolution per minute
OTR	oxygen transfer rate (mmol l^{-1} h^{-1})
OUR	oxygen uptake rate (mmol l^{-1} h^{-1})
P/V	power input (W l^{-1})
p_{O_2}	dissolved oxygen (bar)
q_{CO_2}	specific rate of CO_2 release (in moles g^{-1} h^{-1})
q_{O_2}	specific rate of O_2 uptake (in moles g^{-1} h^{-1})
q_p	specific product formation rate (eg. g g^{-1} h^{-1})
q_s	specific uptake rate of substrate (mmoles l^{-1} h^{-1}; g l^{-1} h^{-1})
RQ	q_{CO_2}/q_{O_2}
r_x	productivity (g g^{-1} h^{-1})
s	substrate (g l^{-1})
s_0	initial substrate concentration (g l^{-1})
SCP	single cell protein
t	time
T	temperature
μ	specific growth rate (h^{-1})
μ_{max}	maximal specific growth rate
x	biomass (g l^{-1})
Y	yield (—)

Abbreviations for Dimensionless Numbers

Ga	Galilei number
Ne	Newton number
Re	Reynolds number

Reactor Configurations

COLOR	compact loop reactor
FBT	flat blade turbine
JLR	jet loop reactor
PLR	propeller loop reactor
SLR	short loop reactor

STR	stirred tank reactor
TORUS	annular configuration
TR	tower form reactor
TLR	tall loop reactor

7 References

1. Finn, R. K.: Bacteriol. Rev. *18*, 255 (1954)
2. Tsao, G.: Biotech. Bioeng. *12*, 51 (1970)
3. Calderbank, B.: Biochem. and Biolog. Eng. Science (Blakebrough, N. ed.), New York: Academic Press 1967
4. Uhl, V. W., Gray, J. B. (eds.): Mixing Theory and Practice, New York: Academic Press 1967
5. Schuegerl, K.: Chem. Ing. Techn. *52* (12), 951 (1980)
6. Taguchi, H.: Adv. Biochem. Eng. *1*, 1 (1971)
7. Brauer, H.: Adv. Biochem. Eng. **13**, 87 (1979)
8. Yoshida, F.: Ind. Eng. Chem. *52*, 435 (1960)
9. Cooper, C. M., Fernstrom, G. A., Miller, S. A.: Ind. Eng. Chem. *36*, 504 (1944)
10. Calderbank, P. H.: Trans. Inst. Chem. Eng. *37*, 173 (1959)μ ibid *36*, 443 (1958)
11. Zlokarnik, M.: Adv. Biochem. Eng. *11*, 157 (1979)
12. Sinclair, C. G., Brown, D. E.: Biotech. Bioeng. *12*, 1001 (1970)
13. Meier, C., Beyeler, W.: Biotech. Bioeng. (in press 1984)
14. Fan, L. T. et al.: Biotech. Bioeng. **12**, 1019 (1970)
15. Charles, M.: Adv. Biochem. Eng. *8*, 1 (1978)
16. Atkinson, B., Daoud, I. S.: Adv. Biochem. Eng. *4*, 41 (1976)
17. Atkinson, B., Fowler, H. W.: Adv. Biochem. Eng. *3*, 221 (1973)
18. Schumpe, A., Quicker, G., Deckwer, W.-D.: Adv. Biochem. Eng. *24*, 1 (1982)
19. Swallow, B., Finn, R. K., Einsele, A.: Proc. 1st Europ. Congr. Biotech. Interlaken 1978
20. Gschwend, K. W.: Stofftransport in nicht-newtonschen Biosystemen. Diss. ETH Nr. 6975, Zurich 1982
21. Sittig, W., Heine, H.: Fortschritte der Verfahrenstechnik *14*, 354 (1977); *11*, 12 (1977)
22. Schuegerl, K.: Verwendung von Blasensaeulen als Bioreaktoren, Lafferty R. L. (ed.) Rothenburger Symp. Ferm. Technik, Braun, Melsungen 1978
23. Schuegerl, K.: Chem. Ing. Tech. *55* (2), 123 (1983)
24. Schuegerl, K., Luecke, J., Oels, U.: Adv. Biochem. Eng. *7*, 1 (1977)
25. Schuegerl, K., Luecke, J., Lehmann, J., Wagner, F.: Adv. Biochem. Eng. *8*, 63 (1978)
26. Blenke, H.: Adv. Biochem. Eng. *13*, 121 (1979)
27. Fiechter, A.: Chem. Ing. Techn. *34* (10), 692 (1962)
28. Einsele, A., Fiechter, A.: Path. E. Microbiol. (Basel) *34*, 149 (1969)
29. Karrer, D.: Der total gefuellte Bioreaktor. Diss. ETH Nr. 6254, Zurich 1978
30. Laederach, H. E.: Ueber fluiddynamische Untersuchungen im Torusbioreaktor. Diss. ETH Nr. 6410, Zurich 1979
31. Herzog, P.: Fluiddynamik und Waermeaustausch Newtonscher und nicht-Newtonscher Gas-Fluessigdispersionen im Torus bioreaktor. Diss. ETH Nr. 7137 Zurich 1982
32. Einsele, A.: Dechema Monographien 157, 1978
33. Wang, D. I. C., Humphrey, A. E.: Chem. Eng. *26*, 108 (1969)
34. Monod, J.: Recherche sur la Croissance des Cultures Bacteriennes. Hermann, Paris 1940
35. Pirt, S. J.: Principles of Microbe and Cell Cultivation, London 1975
36. Janshekar, H.: Studies on Continuous Production of Fodder Yeast from Molasses, Diss. ETH Nr. 6402, Zurich 1979
37. Kaetterer, L.: Wachstumskinetische Untersuchungen an Hefepopulationen von Candida tropicalis und Trichosporon cutaneum. Diss. ETH Nr. 7172, Zurich 1982
38. Adler, I., Fiechter, A.: Chem. Ing. Techn. *4*, 322 (1983)
39. Heer, B., Braendli, E., Fiechter, A.: Regulation of Molatdehydrogenase Isoenzyme in the Glucose-sensitive Yeast Schizosaccharomyces pombe. 5th Int. Ferm. Symp. Berlin 1976

40. Braendli, E.: Untersuchung zur Regulation und Funktion der Malat-Dehydrogenase in Schizosaccharomyces pombe. Diss. ETH Nr. 6549, Zurich 1980
41. Dostalek, M. et al.: Fermentation Technology Today, in: Proc. Int. Fermentation Symp. (Terui, G. ed.) Japan: Soc. Ferment. Technol. 1972
42. Haeggstroem, L.: Appl. Environ. Microbial. *33*, 567 (1977)
43. Mateles, R. I., Battat, E.: Appl. Microbiol. *28*, 901 (1974)
44. Kuhn, H. et al.: Europ. J. Appl. Microbiol. Technol. *6*, 341 (1974)
45. Goldberg, I., Er-el, Z.: Process Biochemistry *16*, 2 (1981)
46. Mateles, R. J.: Biotech. Bioeng. *13*, 581 (1971)
47. Abbott, B. J., Clamen, A.: Biotech. Bioeng. *15*, 117 (1973)
48. Koser, S. A.: Vitamin Requirements of Bacteria and Yeasts. Springfield 1980
49. Suomalainen, H., Oura, E.: Yeast Nutrition and Solute Uptake, in: The Yeasts (Rose, A. H., Harrison, J. S. eds.), p. 3. London 1971
50. Praeve, P., Faust, U., Sittig, W., Sukatsch, D. A.: Handbuch der Biotechnologie. Wiesbaden 1982
51. Hutner, S. H.: Ann. Rev. Microbiol. *26*, 313, 1972
52. Clark, D., Meyer, H.-P., Leist, C., Fiechter, A.: in preparation 1984
Reiling, A., Laurila, H., Fiechter, A.: in preparation 1984
53. Rieger, M.: Untersuchung zur Regulation von Glykolyse und Atmung in Saccharomyces cerevisiae. Diss. ETH Nr. 7264, Zurich 1983
54. Puhar, E., Karrer, D., Einsele, A., Fiechter, A.: 1st Europ. Congr. on Biotechn. Part 2, p. 83, Interlaken 1978
55. Buchholz, H., Luttmann, R., Zakrzewski, W., Schuegerl, K.: Europ. J. Appl. Microb. Biotech. *12*, 63 (1981)
56. Fiechter, A.: Untersuchungen an wachsenden Zellen von Saccharomyces cerevisiae. Habilitationsschrift ETH, Zurich 1966
57. Einsele, A., Fiechter, A.: 3rd Symp. Technische Microbiologie (Dellweg, H. ed.), p. 41, Berlin 1973
58. Karrer, D., Puhar, E., Fiechter, A.: 1st Europ. Congr. on Biotechn. Part 1, p. 23, Interlaken 1978
59. Fiechter, A.: Dechema Monographien. Biotechnology. Proc. 1st Europ. Congress on Biotechn. Survey Lectures Bd. 82, p. 17, 1978
60. Einsele, A., Karrer, D.: Europ. J. Appl. Microb. Biotechn. *9*, 83, 1980
61. Mueller, H.: Chem. Anlagen Verfahren, Nr. 8, 41, 1971
62. Wania, H., Reuss, M., Wagner, F.: 3rd Symp. Technische Mikrobiologie (Dellweg, H. ed.) p. 41, 1973
63. Keitel, G.: Thesis, Univ. Dortmund 1978
64. Koenig, B.: Thesis, Univ. Hannover 1980
65. Seipenbusch, R., Blenke, H.: Adv. Biochem. Eng. *15*, 1, 1980
66. Sittig, W., Heine, H.: Chem. Ing. Tech. *49*, 595, 1977
67. Faust, U., Sittig, W.: Adv. Biochem. Eng. *17*, 63, 1980
68. Laederach, H., Widmer, F., Einsele, A.: Proc. 1st Europ. Congr. on Biotechn. Part I, p. 84, Interlaken 1978
69. Ault, R. G.: Proc. 10th Congr. Europ. Brew. Convention Stockholm, p. 238, Elsevier, Amsterdam 1965
70. Luecke, J., Oels, U., Schuegerl, K.: Chem. Ing. Techn. *48*, 573, 1976
71. Luecke, J., Oels, U., Schuegerl, K.: Chem. Ing. Techn. *49*, 161, 1977
72. Schuegerl, K.: Chem. Ing. Techn. *49*, 605, 1977
73. Burhholz, H., Buchholz, R., Niebeschuetz, H., Schuegerl, K.: Europ. J. Appl. Microb. Biotech. *6*, 115, 1978
74. Buchholz, R., Adler, I., Schuegerl, K.: Europ. J. Appl. Microb. Biotech. *7*, 135, 1979; *7*, 241, 1979; *7*, 333, 1979
75. Steel, R., Miller, T. L.: Adv. Appl. Microb. *12*, 153, 1970
76. Zlokarnik, M.: Ruehrtechnik. Ullmanns Encyklopaedie der Techn. Chemie Vol. 2., 4th Ed., Verlag Chemie Weinheim 1972
77. Knoepfel, H.-P.: Zum Crabtree-Effekt bei Saccharomyces cerevisiae und Candida tropicalis. Diss. ETH Nr. 4906, Zurich 1972

78. Hassan, I. T. M., Robinson, C. W.: Canad. J. Chem. Engin. *58*, 198 (1980)
79. Botton, R., Cosserat, D., Charpentier, J. C.: Chem. Eng. Sci. *35*, 82 (1980)
80. Zlokarnik, M.: Adv. Biochem. Eng. *8*, 133 (1978)
81. Adler, I., Schuegerl, K.: Biotech. Bioeng. *25*, 417 (1983)
82. Margaritis, A., Sheppard, J. D.: Biotech. Bioeng. *23*, 2117 (1981)
83. Serieys, M., Goma, G., Durand, G.: Biotech. Bioeng. *20*, 1393 (1978)
84. Einsele, A., Finn, R. K.: I and EC Proc. Des. Dev. *19*, 600 (1980)
85. Hunt, G., Reismann, H. B., Lago, J.: Chem. Eng. Prog. Symp. Ser. *67*, 108 (1970)
86. Gleiser, I. E., Bauer, S.: Biotech. Bioeng. *23*, 1015 (1981)
87. Esener, A. A., Roels, J. A., Kossen, N. W. F.: Biotech. Bioeng. *23*, 1851 (1981)
88. Huang, S. Y., Chu, W. B.: Biotech. Bioeng. *23*, 1491 (1981)
89. White, J.: Yeast Technology, London 1954
90. Sperber, E.: Arkiv Kemi Mineral. Geol. *21*, A Nr. 3, 1945
91. Dostalek, M., Munk, V., Volfova, O., Fenzl, Z.: Biotech. Bioeng. *10*, 865 (1968)
92. Klug, M. J., Markowetz, A. S.: Biotech. Bioeng. *11*, 427 (1969)
93. Imada, Y., Yamada, K.: Agr. Biol. Chem. *35*, 18 (1971)
94. Moo-Young, M., Shimizu, T., Whitworth, D. A.: Biotech. Bioeng. *13*, 741 (1971)
95. Yoshida, F., Yamane, T., Yagi, H.: Biotech. Bioeng. *13*, 473 (1971)
96. Goma, G., Pareilleux, A., Durand, G.: Arch. Microbiol. *88*, 97 (1973)
97. Hug, H., Blanch, H. W., Fiechter, A.: Biotech. Bioeng. *16*, 965 (1974)
98. Velankar, S. K., Barnett, S. M., Houston, C. W., Thomson, A. R.: Biotech. Bioeng. *17*, 24 (1975)
99. Shafovostona, L. D., Pavlasova, E., Steiskalov, E., Masnerova, E., Begal, V., Slhabanova-Nechichova, J. S.: Mikrobiologija *44*, 5, 863 (1976)
100. Shiloach, J., Bauer, S.: Biotech. Bioeng. *17*, 227 (1975)
101. Monod, J., Cohen-Bazier, G., Cohn, M.: BBA *7*, 4, 585 (1951)
102. Landwall, P., Holme, T.: J. Gen. Microbiol. *103*, 345 (1977)
103. Kovac, L., Berta, F., Psenak, M., Slezaritova, V.: Folia Microbiol. *11* (4), 263 (1966)
104. Aiba, S., Nagai, S., Nishizawa, Y.: Biotech. Bioeng. *18*, 1001 (1976)
105. Slezak, J., Sikyta, B.: Folia Microbiol. *12* (5), 441 (1967)
106. Shebata, T. E., Marr, A. G.: J. Bacteriol. *107*, 210 (1971)
107. Matny, T. S., Sail, J. C.: J. Bacteriol. *92* (4), 960 (1966)
108. Kitai, A., Yamagata, T.: Process Biochem. *5*, 11 (1970)
109. Mori, H., Yano, T., Kobayashi, T., Shimizu, S.: J. Chem. Eng. Jap. *12* (4), 313 (1979)
110. Yano, T. et al.: J. Ferm. Techn. *58*, 259 (1980)
111. Harvey, R. J.: J. Bacteriol. *104*, 698 (1970)
112. Clark, D. J., Maaloe, O.: J. Mol. Biol. *23*, 99 (1967)
113. Herbert, D., Elsworth, R., Telling, R. C.: J. Gen. Microb. *14*, 601 (1956)
114. Altman, P. L., Dittmer, D. S.: Biological Handbook. Washington 1968
115. Swings, J., De Ley, J.: Bact. Rev. *41*, 1 (1977)
116. Dykhuizen, D. E., Hartl, D. L.: Microb. Rev. *47*, 150 (1983)
117. Hoefle, M. G.: Appl. Environ. Microb. *46*, 1045 (1983)
118. Aiba, S., Imanaka, T., Tsunekawa, H.: Biotechnol. Lett. *2* (12), 525 (1981)
119. Reese, E. T. et al.: J. Bact. *59*, 485 (1950)
120. Erickson, K. E. et al.: Biotech. Bioeng. *17*, 327 (1975)
121. Janshekar, H. et al.: Arch. Microbiol. *132*, 14 (1982)
122. Toyama, N., Ogawa, K.: Proc. Symp. Bioconv. Cellulosic Subst. IIT Delhi, p. 305 (1978)
123. Perlozar, M. J. et al.: Arch. Biochem. *25*, 449 (1950)
124. Chahal, D. S. et al.: Dev. Ind. Microbiol. *18*, 433 (1977)
125. Herzog, P., Gschwend, K., Widmer, F., Fiechter, A.: Chem.-Ing. Tech. *55*, 566 (1983)
126. Schuegerl, K.: Adv. Biochem. Eng. *22*, 94 (1982)
127. Zabriskie, D. W., Armiger, W. B., Phillips, D. H., Albano, P. A.: Traders Guide to Fermentation Media Formulation. Traders Protein Division of Traders Oil Mill, Fort Worth 1980

Analyses of Various Control Schemes for Continuous Bioreactors

Pramod Agrawal and Henry C. Lim
School of Chemical Engineering, Purdue University, West Lafayette, Indiana 47907, U.S.A.

A common framework is laid through which the feasibilities and efficacies of various control schemes for continuous bioreactors can be evaluated using only the steady state information. It is shown that many important and practical conclusion can be drawn based on the steady state growth models. For this purpose two well known steady state growth models are used, the Monod model and the substrate inhibition model. The control schemes that are reviewed and theoretically analyzed in terms of potential advantages as well as disadvantages, include turbidostats, nutristats, pH-auxostats and those based on various rates such as the base addition rate, the oxygen absorption rate, the oxygen uptake rate, and carbon dioxide evaluation rate. The feasibility of these control schemes is tested by cheking the local controllability and/or stability criteria, while the practical effectiveness is evaluated by analyzing the steady state gains. Existence of input multiplicity is also checked to point out potentially poor static and/or dynamic performances. Finally, a new control scheme is proposed, which is superior to the conventional continuous bioreactor operations and which allows for a multivariable control scheme.

1 Introduction

Slow dynamics usually associated with microbial growth in bioreactors, can result in off specification products over a significant time period. Unforseen disturbances in a bioreactor can cause havoc with the control system, perhaps resulting in a failure of the reactor and requiring a new start-up procedure. Due to the increasing cost of raw materials and capital investments, failures in biological processes could have catastrophic economic consequences. Therefore, there is a strong economic incentive for proper control of continuous bioreactors.

A recent development in biotechnology is the computerization of laboratory and industrial scale fermentation processes [1]. Real-time digital computers already in use, are suitable for on-line control of bioreactors. Application of computers [2] is however currently limited mostly to data logging and control of indirect variables, such as dissolved oxygen, pH, and temperature. Therefore, a large fraction of computer time is already available for implementation of sophisticated control schemes.

For large scale commerical processes such as cell mass and primary metabolite productions as well as laboratory scale operation for modeling microbial growth, there is a strong incentive to develop efficient control schemes that would enable rapid startup and stabilization of steady states in continuous bioreactors.

Even though numerous control schemes have been proposed for continuous bioreactors no rigorous analyses nor relative efficacies of these control schemes have been reported.

Different modes of controlling continuous bioreactors, that have appeared in literature are: turbidostats [3-11], nutristats [9], pH-auxostats [8,12-16], and those based on the oxygen absorption rate [17-19], the oxygen uptake rate [19], and the carbon dioxide evolution rate [20,12]. In these control schemes a representative variable, which characterizes the steady state balanced growth [22,23] of microorganisms in a chemostat, is controlled by manipulating the flow of nutrient medium to a culture vessel.

Lack of accurate dynamic models has prevented analysis of dynamic responses of control systems and even when a reasonable dynamic model is available the analysis has been based on extensive simulation and the results have been specific to the particular model used. In this work we lay a common framework through which feasibility and efficacy analyses of various control schemes can be made using only the steady state information and show that important and practical conclusions can be drawn based on the steady state growth models. When appropriate, we cite previous simulation or experimental findings. The present work makes no attempt to simulate the dynamic characteristics of continuous cultures under various control schemes. In particular, various control schemes are reviewed and theoretically analyzed in terms of

a) local controllability,
b) local stability,
c) the steady state gain, and
d) input multiplicity.

The feasibility of control schemes is tested by checking local controllability and/or local stability criteria and the practical effectiveness is evaluated by analyzing the steady state gain.

Two well-known steady state growth models used in this analysis are, the Monod [24] model, and the substrate inhibition model [9, 25, 26].

A new control scheme is proposed. Compared to control schemes for the conventional bioreactor operation, the proposed control scheme a) statisfies the local controllability criterion at all steady states, b) can have high steady state gains at open-loop stable steady states, and c) can be made free from any problem associated with input multiplicity. Also, the new method provides additional flexibility to control continuous bioreactor operation by employing a multivariable control scheme.

2 Continuous Cultures

The steady state models used in this work are listed in Table 1, and also listed are the parameter values for each model.

Table 1. List of models and associated parameter values

Model	Form	Parameter values
1. Monod [24]	$\mu = \dfrac{\mu_m s}{K_m + s}$	$\mu_m = 0.4\ h^{-1}$, $K_m = 0.05\ g\ l^{-1}$ $s_F = 1\ g\ l^{-1}$, $y = 0.4\ g\ g^{-1}$ $D \le \dfrac{\mu_m s_F}{K_m + s_F} = 0.381\ h^{-1}$
2. (a) Substrate inhibition [9] Constant yield (b) Variable yield	$\mu = \dfrac{\mu_m s}{K_S + s + s^2/K_i}$ $y = y_0$ $\dfrac{1}{y} = \dfrac{1}{y} + \dfrac{m}{\mu}$	$\mu_m = 0.53\ h^{-1}$, $K_S = 0.12\%\ w/v$ $K_i = 2.2\%\ w/v$, $y_0 = 0.4\ g\ g^{-1}$ $s_F = 4\%\ w/v$ $m = 0.01\ h^{-1}$
3. Substrate Inhibition model with variable yield [11]	$\mu = \dfrac{\mu_m s(1 - as)}{K_s + s + s^2/K_i}$ $y = \dfrac{y_0(1 - as)}{1 + bs + cs^2}$	$\mu_m = 0.504\ h^{-1}$ $a = 0.204\%\ w/v$, $K_3 = 8.49 \times 10^{-4}\%\ w/v$ $K_i = 2.46$ $y_0 = 0.383\ g\ g^{-1}$, $b = 2.96\ (\%\ w/v)^{-1}$ $c = -0.501\ (\%\ w/v)^{-2}$ $s_F = 1\%\ w/v$

2.1 Open-loop Operation: Chemostat

A chemostat, described by Novick and Szilard [27,28] and Monod [29], is a very popular mode of operating a continuous biological reactor. In this open-loop mode, the flow rate of nutrient to the bioreactor is held constant so that the dilution rate is less than the maximum specific growth rate. The medium contains excess of all but one nutrients. A steady state results when the specific growth rate of microorganisms balances exactly with the dilution rate. The steady state is said to be asymptotically stable if the reactor returns to its steady state after small perturbations. However, the attainment and maintenance of the desired steady state may be difficult. This problem, although manifested by many microbial cultures, can be predicted theoretically only for those cultures that exhibit substrate inhibited growth [25,30–31], but not [32–34] for those that follow the Monod [24] or the Monod-Herbert [35] type growth. This is not too surprising, if we consider the limited applicability [36–38] of steady state growth models under dynamic situations.

A continuous culture in its simplest form may be represented by the concentration of substrate, s, and the cell mass concentration, x. The material balances on the cell mass and the substrate of a constant-volume continuous-flow reactor, are

$$dx/dt = (\mu(s) - D)\,x \tag{1}$$

and

$$\frac{ds}{dt} = D(s_F - s) - \frac{\mu(s)}{y(s)}\,x \tag{2}$$

where μ, y, D, and s_F represent the specific growth rate, the yield of cell mass, the dilution rate, and the feed substrate concentration, respectively.

In a vector form Eqs. (1) and (2) can be written as

$$dX/dt = f(X, D) \tag{3}$$

where

$$X = \begin{bmatrix} x \\ s \end{bmatrix} \tag{4}$$

and

$$f(X, D) = \begin{bmatrix} f_1(X, D) \\ f_2(X, D) \end{bmatrix} = \begin{bmatrix} (\mu - D)\,x \\ D(s_F - s) - \dfrac{\mu x}{y} \end{bmatrix} \tag{5}$$

2.2 Closed-loop Operation

Consider the problem of microbial cell production in a continuous bioreactor (Eqs. (1) and (2)). Now let us suppose that we want to design a scheme for controlling the effluent cell concentration, x(t) to a desired values, x_d by manipulating the dilution rate D(t).

There are two problems, the startup problem of driving x(t) from an initial condition x_0 to the set point x_d and also the problem of maintaining x(t) at x_d. The startup problem, although important, is not considered here because the startup period usually represents only a small fraction of the total operational period of a continuous bioreactor. In this work, emphasis is laid on the control of continuous bioreactor at a desired steady state.

A common feedback controller is the proportional-integral-derivative (PID) controller given by

$$0 \leqq D(t) = D_s + K_c\left[(x_d - x) + \frac{1}{\tau_I}\int_0^t (x_d - x)\,dt' + \tau_d \frac{d(x_d - x)}{dt}\right] \tag{6}$$

where D_s is the steady state operating value of dilution rate, and K_c, τ_I and τ_d are controller parameters. For simplicity the proportional control mode will be used to compare various control schemes.

3 Criteria for Control System Evaluation

3.1 Controllability

Prior to tackling a design of control system, a notion called controllability is very useful in analyzing control systems. A precise definition of controllability is given in Refs. [39,40], but loosely speaking, a system is said to be controllable if there exists a control policy $U(t)$ which will drive the system from any given initial state X_0 to any other desired state X_d in finite time.

Consider a linear system

$$dX/dt = AX + BU \tag{7}$$

$$Y = CX \tag{8}$$

where X is a state vector of dimension n, U is a control vector of dimension m, Y is an output vector of dimensional l, and A, B, and C are constant matrices of dimensions $n \times n$, $m \times n$, and $l \times n$, respectively. This system is said to be completely controllable if and only if the rank of an $n \times nm$ "controllability matrix" L_c is n, where

$$L_c \equiv [B|AB|A^2B|\ldots|A^{n-1}B] \tag{9}$$

The notion of output controllability, which applies to the output Y, means the outputs are controllable if and only if the rank of the l × nm controllability matrix L_c^0 is l, where

$$L_c^0 \equiv [CB|CAB|...|CA^{n-1}B] \tag{10}$$

Unfortunately, a rigorous controllability criterion can be derived only for the linear system described above. However, local controllability criterion can be obtained through linearization of the nonlinear Eqs. (3) to (5) and application of the linear theory to the linearized equations. Local controllability is necessary, but not sufficient, for controllability of the original nonlinear Eqs. (3) to (5). Nevertheless, for most nonlinear systems good practical conclusions can be drawn by checking the local controllability criterion.

Linearizing Eqs. (3) to (5) around a non-trivial ($x \neq 0$) steady state

$$A = \partial f/\partial X|_{s.s.} = \begin{bmatrix} 0 & \mu' x \\ -\dfrac{\mu}{y} & -\mu - \dfrac{\mu' x}{y} + \dfrac{\mu y' x}{y^2} \end{bmatrix}_{s.s.} \tag{11}$$

where the prime denotes derivative with respect to s, also,

$$b = \partial f/\partial D|_{s.s.} = \begin{bmatrix} -x \\ s_F - s \end{bmatrix}_{s.s.} = \begin{bmatrix} -x \\ x/y \end{bmatrix}_{s.s.} \tag{12}$$

and the local controllability matrix

$$L_c = [b|Ab]_{s.s.} = \begin{bmatrix} -x & \mu' x^2/y \\ x/y & -\mu' x^2/y^2 + \mu y' x^2/y^3 \end{bmatrix}_{s.s.} \tag{13}$$

The system of Eqs. (3) to (5) satisfies the local controllability criterion if and only if the rank of L_c is two, or det $L_c \neq 0$, where

$$\det L_c = \frac{\mu y' x^3}{y^3} \tag{14}$$

For steady states of practical interest $x \neq 0$ and μ and y are non-zero quantities. Thus, L_c is singular only when $y' = 0$. Therefore, it can be concluded that the system of Eqs. (3) to (5) is locally controllable at all the non-trivial steady states, except at a steady state where the yield coefficient is constant or goes through local minima or maxima, $dy/ds = 0$.

3.2 Stabilizability [39,40]

A system is said to be stabilizable if the unstable states of the system can be made stable by controller action. Stabilizability condition for a system is much weaker than the controllability condition. A system which is controllable is automatically stabilizable.

The linear system of Eq. (7) is stabilizable, if the real parts of every eigenvalue of A can be made negative by controller action. If the real parts of every eigenvalue of A is negative, the system is automatically stabilizable (even without controller action).

Consider a proportional controller fed back on the state variables

$$U(t) = -KX(t) \tag{15}$$

where K is an $m \times n$ constant gain matrix, then Eq. (7) becomes

$$dX/dt = (A - BK)\,X \tag{16}$$

If elements of K can be chosen such that the real parts of all the eigenvalues of (A — BK) are negative, then the system is stabilizable by the above feed back controller.

With a proportional controller fed back on the output variables (Eq. (8)), we have

$$U(t) = -KCY(t) \tag{17}$$

and

$$dX/dt = (A - BKC)\,X \tag{18}$$

The output stabilizability requires that the real parts of all the eigenvalues of (A — BKC) be negative for some choice of the feed back gain matrix K.

For nonlinear systems local stabilizability conditions can be obtained by linearizing the system about a steady state and applying the results obtained for linear systems.

From Eq. (14) we conclude that a microbial system that exhibits a constant cell-mass yield ($y' = 0$) is not completely controllable in a continuous bioreactor. However, by manipulating D, it is possible to control x or s, but not both.

For the case of constant yield $y' = 0$, let

$$z = x - y(s_F - s) \tag{19}$$

Equations (1) and (2) can be reduced to

$$\frac{dx}{dt} = (\mu - D)\,x \tag{1}$$

and

$$\frac{dz}{dt} = -Dz \tag{20}$$

The last equation yields

$$z = z(0)\,e^{-\int_0^t D(t_1)\,dt_1} \tag{21}$$

Since, x can be controlled by manipulating D and the eigenvalue associated with z is always negative (hence z is stable), the system of Eqs. (1) and (20) (or (1) and (2)) is stabilizable. Even though the system is not completely controllable, it is possible to design a successful control system in which x is controlled and would take on the value given by $s = s_F + x/y$.

This type of continuous bioreactor control scheme is termed a turbidostat. Similarly, it is possible to design a control system in which s is controlled by manipulating D and x automatically reaches a steady state. Nutristat is a name given to this control scheme.

In general controllability and stabilizability have the following implications [40] in the design of control systems. If a system is completely controllable, a control scheme can always be designed. It may be impossible to design a control scheme for a system which is not completely controllable. But if such a system is stabilizable and the uncontrollable eigenvalues of the system are sufficiently large and negative, then it is possible to design an acceptable control scheme. However, it would generally be impossible to design a control scheme for a system which is not stabilizable.

3.3 Steady State Process Gain

The steady state process gain, K_y, is defined as the change in the controlled (output) variable resulting from a unit change in the manipulated variable, dY/dD. For a (scalar) proportional control, the proportional gain K_c is inversely proportional to the steady state process gain K_y, since the ultimate gain, K_cK_y, is the system property and may be assumed to be constant.

The steady state gain, K_y, is an important parameter in determining the effectiveness of a control scheme. A small value of K_y indicates an insensitivity of the manipulated variable to changes in the controlled variable. For example, a small value of K_y implies that a large change in the manipulated variable (controller action) is needed to correct for a small change in the controlled variable. This can cause saturation in the manipulated variable and can result in poor control action. Also, if there is a small and constant error in the measurement or estimation of the controlled variable, the manipulated variable, because of low K_y, would settle at a value entirely different from its designed steady state value. On the other hand, a high value of K_y implies that a small change in the manipulated variable is needed to correct for the controlled variable, thus avoiding the problem of potential valve saturation and poor control. Consequently, the steady state gain plays an important role in controller tuning.

3.4 Input Multiplicity [41]

A control system has multiple steady states when more than one set of manipulated (input) variables *U* can produce the desired steady state outputs *Y*. This type of multiplicity in a closed-loop system is called the input multiplicity. This is to distinguish from the output multiplicity in open-loop systems where one value of input *U* could produce more than one set of steady state values *Y*. In the presence of input multiplicity, a control system may exhibit poor static performance or dynamic performance, or both. The static performance suffers from the fact that the controlled system may find an alternate steady state (since there is more than one value of the manipulated variable which maintain the same controlled variable) instead of finding the designed. The dynamic performance suffers due to poor controllability in the presence of multiple steady states. The necessary condition [41] for existence of input multiplicity in a nonlinear multivariable control system is the singularity of process gain matrix $\partial Y/\partial U$ at some steady states, or

$$\det(\partial Y/\partial U) = 0 \tag{22}$$

For a scalar control system Eq. (22) reduces to

$$K_y = \partial Y/\partial U = 0 \tag{23}$$

Equation (22) is also called the catastrophe condition [41].

We shall now analyze and compare various control schemes in terms of local controllability, local stabilizability, steady state gain, and input multiplicity.

4 Continuous Culture Control Schemes

4.1 Turbidostat

The evolution of laboratory scale chemostat and turbidostat [3] took place at the same time. In turbidostat the optical density, which is related to the cell mass concentration, is controlled by manipulating the medium flow rate. At a steady state the optical density or the turbidity in the bioreactor is held constant and hence the name "turbidostat". Turbidostat was introduced for a variety of reasons. Unlike a chemostat in which the growth is nutrient limited, the turbidostat provides opportunities to study growth under excess of nutrients [3]. It is also used to study the selection of mutants surviving in excess, toxic nutrient environments [5]. A turbidostat could be used to operate a bioreactor at near the maximum specific growth rate, a task for which chemostat is found to be unsuitable. It can also be used to study the development of mutants adaptable to higher growth-rate conditions [4]. Zines and Rogers [7] studied the effect of external addition of product inhibitor, ethanol, on the growth of *Klebsiella aerogenes* in a turbidostat. The turbidostat response to the above disturbance was oscillatory. Edwards et al. [9] proposed the turbidostat scheme to control a continuous bioreactor at open-loop stable steady states, when growing microorganisms that exhibit substrate-inhibited kinetics. Whaite and Gray [10] used a turbidostat to maximize the productivity, (μxv), of *Candida utilis*. However, the control was implemented in a feed-forward fashion which partly could be the reason for the process instabilities observed in their work. DiBiasio et al. [11] working with a methanol utilizing bacterium *L3*, which exhibits the substrate-inhibition kinetics, employed turbidostat to study the growth kinetics at open-loop unstable steady states. Natural instability and control of unstable steady states in continuous cultures was demonstrated in their work. Figure 1 presents a representative set of experimental profiles of x and s for growth of *L3* in a turbidostat. Batch growth of *L3* occurred until an elapsed time of 15 hours. Thereafter, a proportional control was used to stabilize x near its set-point value. Within the accuracy of the experimental analysis, s also reached a steady state value. After sufficient time at the steady state the controller was disconnected as indicated by the arrows in Fig. 1 and the dilution rate was left at its steady state value. The washout which began shortly thereafter demonstrated instability of the steady state.

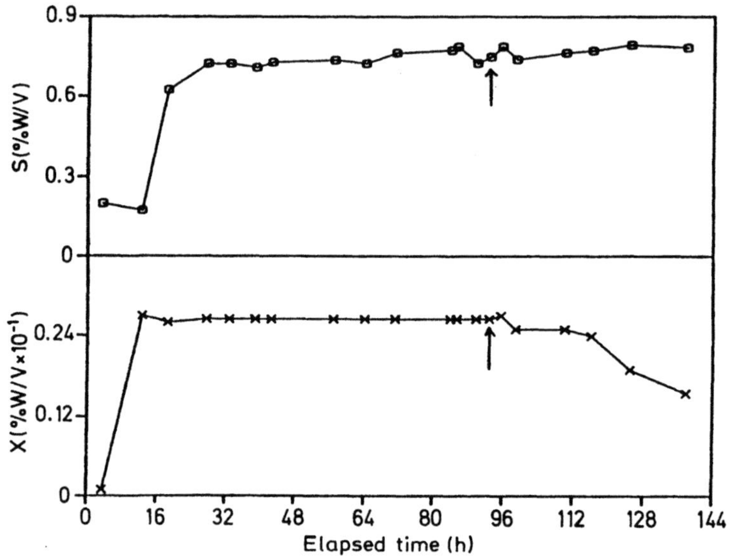

Fig. 1. Demonstration of control of unstable steady state and natural instability in a continuous culture of *L3* (after: DiBiasio et al. [11])

Assuming a proportional control on x in a turbidostat, we have

$$D(t) = D_s + K_{ct}(x_d - x) \geqq 0 \tag{24}$$

where k_{ct} is the proportional controller gain.

With the above proportional controller, the linearized closed-loop system of Eqs. (3)—(5) becomes

$$dX/dt = (A - bK_t^T)\, X \tag{25}$$

where A and b are given by Eqs. (11) and (12), respectively, and

$$K_t^T = [K_{ct}, 0] \tag{26}$$

Thus,

$$(A - bK_t^T) = \begin{bmatrix} K_{ct}x & \mu' x \\ -K_{ct}\frac{x}{y} - \frac{\mu}{y} & -\mu - \frac{\mu'}{y}x + \frac{\mu}{y^2}y'x \end{bmatrix}_{s.s.} \tag{27}$$

For local stability of the closed-loop system, K_{ct} should be chosen such that the real parts of the eigenvalues of $(A - bK_t^T)$ are negative, or,

$$\det(A - bK_t) = \left[\left\{\mu' - K_{ct}\left(y - \frac{y'x}{y}\right)\right\}\frac{\mu x}{y}\right]_{s.s.} > 0 \tag{28}$$

and

$$\mathrm{tr}(A - bK_t^T) = \left[K_{ct}x - \frac{\mu' x}{y} - \frac{\mu}{y}\left(y - \frac{y'x}{y}\right)\right]_{s.s.} < 0 \tag{29}$$

At open-loop steady states, from Eqs. (1) and (2) we obtain

$$x = y(s_F - s) \tag{30}$$

Therefore, the steady-state process gain, $K_y = K_x = dx/dD$ is obtained by differentiating Eq. (30) and recognizing $\mu = D$ at steady states

$$K_X = \frac{dx}{dD} = [(s_F - s)\, y' - y/\mu']_{s.s.}$$

$$= -[(y - y'x/y)/\mu']_{s.s.} \tag{31}$$

Upon substituting Eq. (31) into Eqs. (28) and (29), we obtain

$$\det(A - bK_t^T) = \left[\{\mu'(1 + K_{ct}K_x)\}\frac{\mu x}{y}\right]_{s.s.} > 0 \tag{32}$$

and

$$\mathrm{tr}(A - bK_t^T) = \left[K_{ct}x - \frac{\mu' x}{y} + \frac{\mu\mu'}{y}K_x\right]_{s.s.} < 0 \tag{33}$$

Table 2 gives a typical classification of steady states in continuous bioreactor under turbidostat control. It also indicates the proper range of K_{ct} values for stable closed-loop operation.

For the constant yield case ($y' = 0$) the stability conditions given by Eqs. (28) and (29) reduce to a single condition given by

$$[\mu' - K_{ct}y]_{s.s.} > 0 \tag{34}$$

If $\mu' > 0$, the steady state is open-loop stable, and by choosing $K_{ct} < 0$ the stability characteristics of the closed-loop system can be improved. For $\mu' < 0$ the open-loop steady state is unstable and by choosing K_{ct} such that

$$K_{ct} < \left[\frac{\mu'}{y}\right]_{s.s.} < 0 \tag{35}$$

the steady state can be stabilized.

For the Monod model with a constant cell mass yield the steady state gain K_x is given by

$$K_x = -\frac{\mu_m K_m y}{(\mu_m - D)^2} \tag{36}$$

Table 2. A typical classification of steady state in turbidostat

$\mu' > 0$		$\mu' < 0$
$K_x > 0$	$K_x < 0$	$K_x > 0$
$^*K_{ct} \lesseqgtr 0$	$K_{ct} < 0$	$K_{ct} < \frac{-1}{K_x} < 0$

* Depending on the operating conditions

At low to moderate dilution rates where the growth is substrate limited, the steady state gains are very low and the turbidostat scheme may not be reliable. However, at dilution rates close to the maximum specific growth rate, where the growth is not substrate limited, the cell density falls rapidly with the dilution rate and the steady-state gains are high. Thus, the turbidostat would be effective. For given values of μ_m and s_F, higher the values of $K_m y$ larger will be the effectiveness range of turbidostat operation.

Figure 2 shows plots of (a) the cell concentration x and (b) the steady state gain K_x vs. the dilution rate D for the Monod model (1) and a substrate inhibition model (2a). The low conversion steady state in the substrate inhibition model is open-loop unstable [30,42]. In Fig. 2 and subsequent figures, the solid lines represent the open-loop stable steady states, while the dotted lines represent the open-loop unstable steady states.

For the substrate inhibition model the region of rapidly falling cell concentration does not represent the open-loop stable steady states. Therefore, K_x at stable steady states has low values, except in a narrow range of dilution rates near the maximum specific growth rate. However at the unstable steady states, K_x values are high, making the turbidostat scheme effective as was experimentally demonstrated by DiBiasio et al. [11].

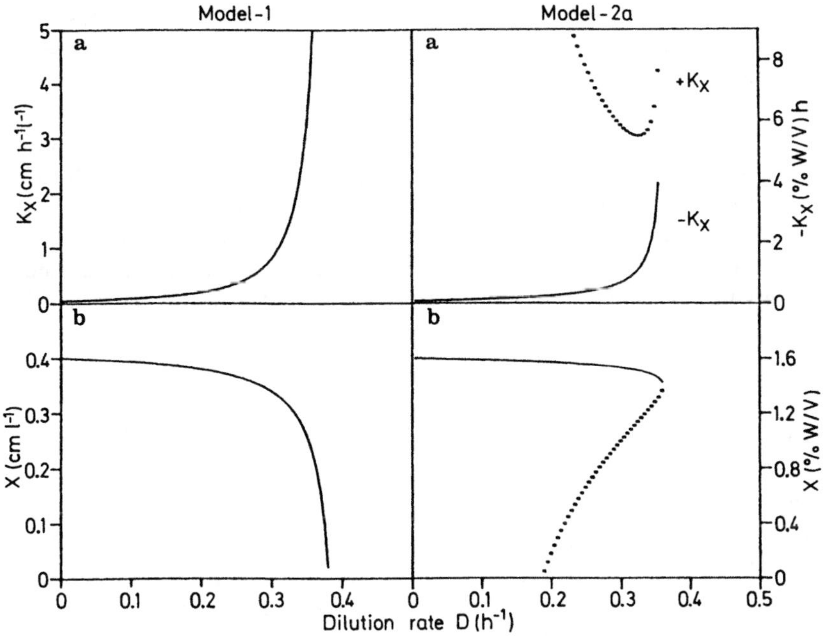

Fig. 2a and b. Plots of (**a**) steady state gain, K_x, and (**b**) cell mass concentration, x, vs. dilution rate, D, for the Monod model (1) and a substrate inhibition model 2a

If the cell mass yield varies as a result of the maintenance requirement (Model 2b), K_x is obtained from Eq. (31) as

$$K_x = \frac{m y_0^2 s_F}{(\mu + m y_0)} - \frac{\mu y_0}{(\mu + m y_0)\,\mu'} \tag{37}$$

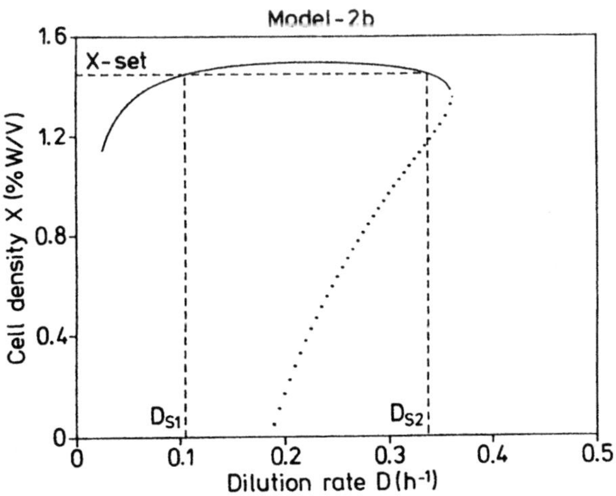

Fig. 3. Plot of cell mass concentration, x, vs. dilution rate, D, for a maintenance model 2b

Thus, K_x changes sign (goes through zero) at an open-loop stable steady state and satisfies Eq. (23). Therefore, the maintenance term may lead to the problems associated with input multiplicity. This is so illustrated in Fig. 3 where two different dilution rates yield the same cell concentration.

The controller in turbidostat scheme is constrained. When $x < x_d$, the controller reduces the nutrient flow rate and therefore, indirectly increases the cell concentration by washing out less cells rather than directly adding cells to the bioreactor. If the integral action is also used in the control algorithm, the error builds up forcing D to be zero in Eq. (6). Furthermore, if the transient cell mass yields are low, it is not difficult to conceive of a situation where $x < x_d$, $s = 0$ and $D = 0$. When this happens, the controller will fail. This is another shortcoming of a turbidostat.

4.2 Nutristat

As early as 1957 Fuld and Dunn [43] proposed automatic control of fermentors based on the residual substrate concentration. Edwards et al. [9] used the term nutristat for such a type of bioreactor control scheme. Unavailability of a measurement device capable of continuously monitoring the residual substrate concentration in a bioreactor, hindered the application of nutristat. Much effort is made by many [44-49] to develop sensors for industrially important substrates. A successful report dealing with direct nutristat control of continuous bioreactors is yet to appear. However, control of nutrients in fed batch fermentors has been reported [43,47].

For a proportional control on the residual substrate concentration in a nutristat, we have

$$D(t) = D_s + K_{cn}(s_d - s) \geqq 0 \tag{38}$$

$$dX/dt = (A - bK_n^T)\,X \tag{39}$$

where K_{cn} is the proportional controller gain, s_d is the set point value of the substrate concentration, and

$$K_n^T = [0 \;\; K_{cn}] \tag{40}$$

Thus,

$$(A - bK_n^T) = \begin{bmatrix} 0 & \mu' x + K_{cn} x \\ -\dfrac{\mu}{y} & -\mu - \dfrac{\mu' x}{y} + \dfrac{\mu y' x}{y^2} - \dfrac{K_{cn} x}{y} \end{bmatrix}_{s.s.} \tag{41}$$

and K_{cn} should be chosen such that

$$\det(A - bK_n^T) = \left[\frac{\mu x}{y}(\mu' + K_{cn})\right]_{s.s.} > 0 \tag{42}$$

and

$$\mathrm{tr}(A - bK_n^T) = \left[- K_{cn}\frac{x}{y} - \frac{\mu' x}{y} - \frac{\mu}{y}\left(y - \frac{y' x}{y}\right)\right]_{s.s.} < 0 \tag{43}$$

Clearly, from Eqs. (42) and (43), a positive K_{cn} should be chosen for a proper nutristat control at all steady states in continuous bioreactor.

For a nutristat the steady state gain K_s is given by

$$K_s = \left[\frac{ds}{dD}\right]_{s.s.} = \frac{1}{[\mu']_{s.s.}} \tag{44}$$

For the case of constant yield ($y' = 0$) K_{cn} should be such that

$$[\mu' + K_{cn}]_{s.s.} = \frac{1}{K_s} + K_{cn} > 0 \tag{45}$$

For control at open-loop stable steady states ($\mu' > 0$), $K_{cn} > 0$; and for stabilization of the open-loop unstable steady states ($\mu' < 0$),

$$K_{cn} > (-\mu')_{s.s.} = -\frac{1}{K_s} > 0 \tag{46}$$

From Eqs. (31) and (44) for constant yield case

$$K_s = -\frac{1}{y}K_x \tag{47}$$

In general the cell mass yield (g of cells per g of substrate) is less than one, therefore, the steady state nutristat gains are higher than the turbidostat gains. Thus, the nutristat response is expected to be better than the turbidostat response, consistent with the extensive simulation results of Edwards et al. [9]. The cell mass yield of most microorganisms is in the neighborhood of 0.5 g g^{-1}, hence the steady state gains of nutristat would be about twice the turbidostat gains. This moderate increase in K_s over K_x may not be sufficient to ensure a feasible nutristat control in the region where turbidostat is ineffective. Also, the substrate concentrations at substrate limited growth conditions are usually very low so that even small measurement errors would adversely affect the nutristat operation. Nutristats, however, have other advantages over turbidostats: a) absence of input multiplicity and b) absence of problems associated with the constrained controller, as found in turbidostats; contrary to turbidostat, a nutristat requires positive controller action ($K_{cn} > 0$, whereas K_{ct} is generally less than zero).

4.3 pH-Auxostat

The pH has a marked influence on the rate of microbial growth. Characteristically there is a pH value at which the growth rate is optimal and on each side of this optimum the growth rate is lower. The pH is controlled by the addition of suitable buffers.

In bioreactors, the pH is usually controlled at an optimum value by the addition of acid or base. Changes in the pH occur as a result of nutrient uptake or due to acid production by microorganisms. Manipulation of the nutrient flow based on the pH change in continuous culture is possible, if there exisits a definite correlation between the microbial growth and the pH change. Watson [8] indicated the possibility of continuous-culture control based on pH as the controlled variable. Martin and Hempling [12] gave a brief analysis of a pH-auxostat, and since then few reports have appeared on successful operations of pH-auxostats [12-17]. Interestingly a pH auxostat was shown by Martin and Hempling [12] to cause a smooth transition of a continuous culture of *E. coli B* from aerobic to anaerobic growth condition and vice versa. Their experimental results are reproduced in Fig. 4. When the steady state aerobic growth

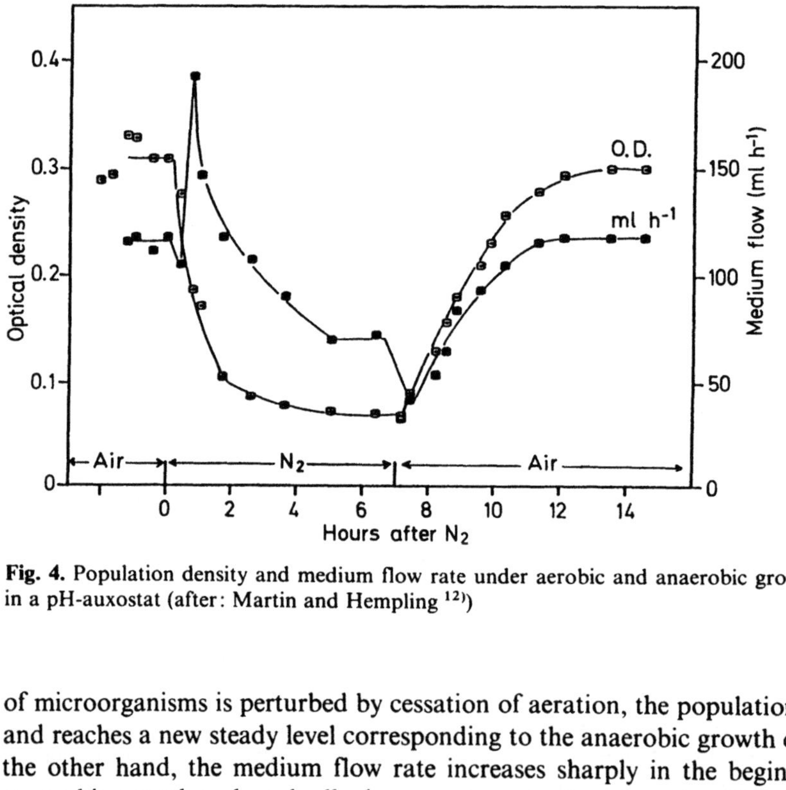

Fig. 4. Population density and medium flow rate under aerobic and anaerobic growth of *E. coli B* in a pH-auxostat (after: Martin and Hempling [12])

of microorganisms is perturbed by cessation of aeration, the population density falls and reaches a new steady level corresponding to the anaerobic growth condition. On the other hand, the medium flow rate increases sharply in the beginning stage of anaerobic growth and gradually decreases to a steady state value. Upon reaeration of the continuous culture at 7 hours the population density increases to the original steady state level corresponding to the aerobic growth condition, while the medium flow rate decreases sharply first and then increases to the original steady state value.

Consider a microbial growth system where the growth is associated with the production of acid (H^+ ions). In pH-auxostat, the pH is controlled by feeding an appropriate amount of medium containing either a buffer or base [13,15]. We define the buffering capacity, BC, of a medium as BC moles of H^+ ions absorbed by a liter of medium to change its pH to the set point pH (pH_d) in the bioreactor. The H^+ ion balance in a continuous-biological reactor yields

$$\frac{d(H^+)}{dt} = \mu x y_{H^+/x} - D(BC) - D(H^+ - H_d^+) \tag{48}$$

where $y_{H^+/x}$ is the yield of acid (moles of H^+ per g of cell mass) and H_d^+ is the set point pH, pH_d. At a steady state, $dH^+/dt = 0$, $H^+ = H_d^+$, and $\mu = D$, and therefore, from Eq. (48) we obtain

$$BC = x y_{H^+/x} \tag{49}$$

Thus, at a controlled steady state the buffering capacity of nutrient medium determines the cell mass concentration x, which in turn determines the steady state value of the specific growth rate.

The steady state gain of pH-auxostat is related to the steady state gain of turbidostat by $y_{H^+/x}$,

$$K_{H^+} = y_{H^+/x} K_x \tag{50}$$

An interesting situation occurs when the specific rate of acid production μ_{H^+}, follows Luedeking and Piret [50] form (Eq. (51)) as in the lactic acid production [16] from lactose under nitrogen or energy limited conditions. The specific rate of acid production is given by

$$\mu_{H^+} = \mu y_{H^+/x} = \alpha\mu + \beta \tag{51}$$

and therefore,

$$y_{H^+/x} = \alpha + \beta/\mu \tag{52}$$

where α and β are constants. The steady state gain K_{H^+} is obtained by substituting Eq. (52) to Eq. (50)

$$K_{H^+} = (\alpha + \beta/\mu)\, K_x \tag{53}$$

The K_{H^+} values are high at low dilution rates ($\mu = D$) and therefore, unlike the turbidostat scheme the pH-auxostat scheme would work well at low dilution rates.

4.4 Rate Controlled Continuous Cultures

4.4.1 Base-Addition Rate (BAR) Controlled Bioreactor

In a pH-auxostat the pH in the bioreactor may vary substantially from the setpoint value during the transient to the desired steady state. Variations in the pH could have deleterious effects on microbial growth [51, 52]. This could make the pH-auxostat undesirable. The pH-auxostat also has another disadvantage due to its operational limit caused by the buffering capacity of the nutrient medium [13,15]. These drawbacks of pH-auxostats suggest a new mode of control in which the pH is independently controlled by the addition of a base, and the base addition rate, (BAR), is controlled by manipulating the dilution rate. A BAR controlled bioreactor is currently under experimental investigation in our laboratory.

For simplicity we assume the pH of the nutrient medium to be the same as the steady-state pH in the bioreactor. The rate of acid production by microorganism is given by

$$vr_{H^+} = \mu_{H^+} vx = \mu y_{H^+/x} vx \tag{54}$$

The pH is maintained constant usually by the addition of a base or an acid. To maintain a constant pH in a continuous culture, the base addition rate must be equal to the acid production rate. Thus,

$$BAR = vr_{OH^-} = vr_{H^+} = y_{H^+/x}(\mu vx) \tag{55}$$

At a controlled steady state the base addition rate determines the cell mass productivity (μvx).

4.4.1.1 Output Controllability

We consider BAR as an output variable for the system of Eqs. (3) to (5), therefore,

$$Y = BAR = y_{H^+/x} \mu vx \tag{56}$$

and for simplicity we assume $y_{H^+/x}$ to be constant. The linearized output controllability matrix L_c^0 is given by Eq. (10), or

$$L_c^{\,0} = [cb \mid cAb]_{s.s.} \tag{57}$$

where A and b are given by Eqs. (11) and (12), respectively, and

$$c^T = \left[\frac{\partial y}{\partial x} \; \frac{\partial y}{\partial s}\right]_{s.s.} = y_{H^+/x} v [\mu \mid \mu' x]_{s.s.} \tag{58}$$

Substituting Eqs. (11), (12) and (58) into Eq. (57) we obtain

$$L_c^0 = y_{H^+/x} v \left[x\left(-\mu + \frac{\mu' x}{y}\right) \middle| \frac{\mu' x^2}{y}\left(\mu - \frac{\mu' x}{y} + \frac{\mu y' x}{y^2}\right)\right]_{s.s.} \tag{59}$$

The output variable BAR is locally controllable, if and only if at least one of the two terms in L_c^0 is non-zero. The two terms are simultaneously zero only when

$$\left[\frac{\mu' x}{y} - \mu\right]_{s.s.} = 0 \quad \text{and} \quad y' = 0 \tag{60}$$

For a microbial system that exhibits a constant cell-mass yield ($y' = 0$) at steady states

$$\frac{d(\mu x v)}{dD} = \frac{vy}{\mu'}\left[\frac{\mu' x}{y} - \mu\right] \tag{61}$$

Thus, the condition given by Eq. (60) states that the output (BAR) fails to satisfy the local controllability criterion only at a steady state that corresponds to the maximum cell mass productivity in a continuous microbial system.

From Eq. (55) the steady state gain, K_b, for a BAR controlled bioreactor is

$$K_b = \frac{d(\mathrm{BAR})}{dD} = y_{H^+/x} v \left[x + D\frac{dx}{dD}\right]_{s.s.} \tag{62}$$

Figure 5 shows the splots of a) K_b and b) BAR vs. the dilution rates for the Monod model (1) and the substrate inhibition model (3). An interesting feature of the rate

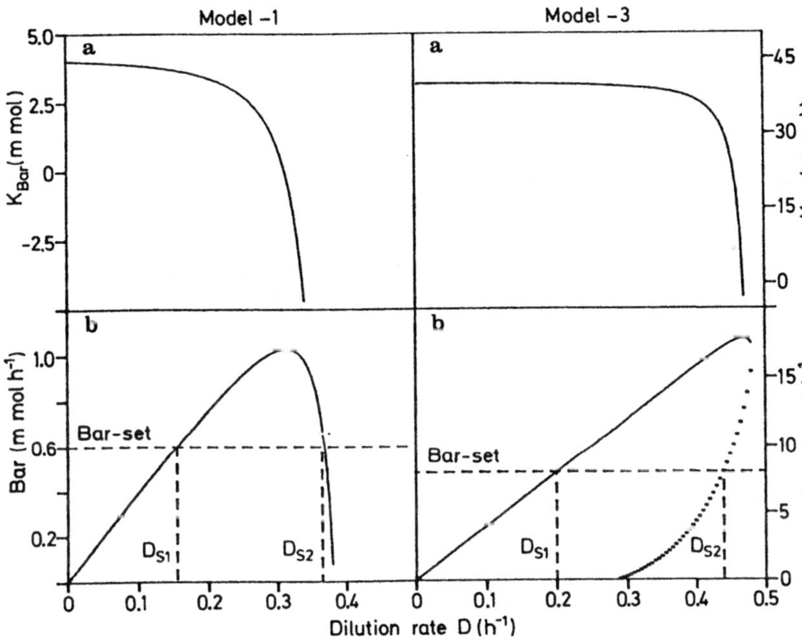

Fig. 5a and b. Plots of (**a**) steady state gain, K_{BAR}, and (**b**) base-addition rate, BAR, vs. dilution rate, D, for the Monod model 1 and a substrate-inhibition model 3. V = 1.0 l and $Y_{H^+/x}$ = 10.0 mmol (OH^-) per g cell-mass

controlled operations, as opposed to the control modes considered previously, is that the steady state gain depends not only on the growth associated term (i.e. term involving dx/dD) but also on xv. Therefore, the steady state gain depends on the operating conditions of a bioreactor, such as s_F and v. As a result, the K_b values are higher than the corresponding K_x values in the substrate limited growth region (where dx/dD is low). In this region a rate controlled operation would be effective.

If a proportional controller is employed in a BAR-controlled bioreactor, we have

$$D(t) = D_s + K_{cb}(BAR_d - BAR) \geqq 0 \tag{63}$$

$$dX/dt = (A - b^T K_{cb} c)\, X \tag{64}$$

where K_{cb} is the controller gain. For local stability of the closed loop system the choice of K_{cb} should be such that the eigenvalues of $(A - b^T K_{cb} c)$ have negative real parts. From Eqs. (11), (12), (58), (62) and (64) we obtain the following stability conditions

$$\det(A - b^T K_{cb} c) = \left[\mu' \left\{1 + \frac{K_{cb}}{y_{H^+/x}} K_b\right\} \frac{\mu x}{y}\right]_{s.s.} > 0 \tag{65}$$

$$\mathrm{tr}(A - b^T K_{cb} c) = \left[\mu - \frac{\mu' x}{y}\left(1 + \frac{K_{cb}}{y_{H^+/x}} K_b\right) + \frac{\mu y' x}{y^2}(1 - K_{cb} v x)\right]_{s.s.} < 0 \tag{66}$$

For constant yield ($y' = 0$) we have a simplified stability condition given by

$$\left[\mu'\left(1 + \frac{K_{cb} K_b}{y_{H^+/x}}\right)\right]_{s.s.} > 0 \tag{67}$$

Table 3 gives a typical classification of steady states in a rate-controlled continuous bioreactor. The table also gives proper K_{cb} limits for a stable closed-loop operation. At the maximum cell mass productivity, $d(\mu x v)/dD = 0$, the steady state gain K_b changes its sign and is equal to zero. Therefore, from Eq. (67) (and also Eq. (60)) control near the maximum cell productivity is very difficult.

Table 3. A typical classification of steady states in the BAR controlled continuous bioreactor

$\mu' > 0$		$\mu' < 0$
$K_b > 0$	$K_b < 0$	$K_b > 0$
$K_{cb} > 0$	$K_{cb} < 0$	$K_{cb} < \frac{-y_{H^+/x}}{K_{cb}} < 0$

The rate-controlled bioreactor operation exhibits input multiplicity because the steady state gain K_b changes sign and satisfies the condition given by Eq. (23). As shown in Fig. 5, for both models (Monod and Substrate Inhibition) there exist two steady state values of dilution rates for a given set point value of the base addition rate (except at the maximum productivity). Let D_{S1} and D_{S2} ($D_{S2} > D_{S1}$) represent the two steady state dilution rates for a set point value of BAR (Fig. 5). In general, for a particular choice of K_{cb}, say $K_{cb} > 0$, the steady state corresponding to the low dilution rate D_{S1} would be stable, while the steady state corresponding to the higher dilution rate D_{S2} would be unstable. Therefore, the input multiplicity would not cause poor static performance of the closed-loop system. Ultimately, the system would settle at the designed steady state.

As opposed to the turbidostat scheme, the BAR-controlled bioreactor operation would be little, if at all, affected by a) variations in the cell-mass yield due to the maintenance requirement, and b) constraints on the controller.

4.4.2 Oxygen Absorption Rate (OAR) and Oxygen Uptake Rate (OUR) Controlled Bioreactors

For aerobic cultures the two most important substrates are the carbon source in the nutrient medium and oxygen in the fermentor gas stream. In these modes of operations, the absorption/uptake rate of oxygen is used to regulate the feed flow rate of the nutrient medium containing the carbon source. The dissolved-oxygen concentration (DOC) which is related to OAR and the off-gas oxygen concentration (OOC) which is related to OUR, can be used as controlled variables in OAR and OUR controlled schemes, respectively. The control based on OAR was used by Hospodka [17] to monitor a continuous culture of *Saccharomyces cerevisiae*. The regulation of pulse feed [18, 53] of the limiting nutrient by a control based on the DOC level has been reported also. Yamada et al. [19] controlled the continuous fermentation of sorbital to sorbose by *Acetobacter suboxydans* using the partial pressure of oxygen in effluent gas and also the dissolved oxygen concentration in culture broth as controlled variables. The control scheme based on the partial pressure of oxygen gave a better performance than the one based on the dissolved oxygen concentration. Their typical experimental results on continuous culture control based on the measure-

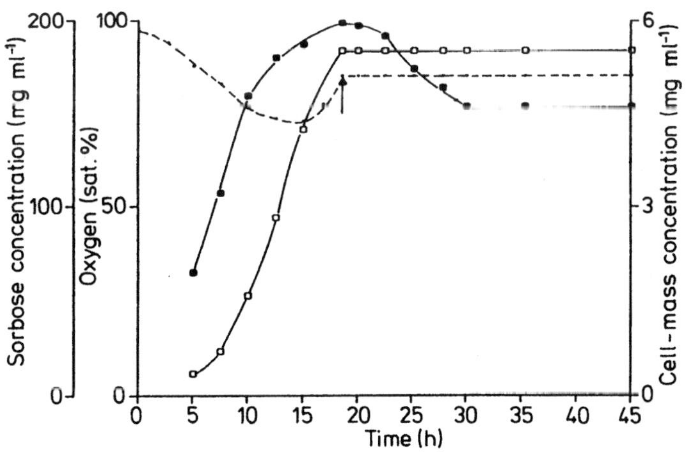

Fig. 6. Time profiles of cell mass and sorbose concentrations and partial pressure of oxygen in effluent gas in a OUR controlled continuous culture of *Acetobacter suboxydans* (after: Yamada et al. [19].) (——— oxygen, —□— sorbose, and —■— cell mass)

ment of partial pressure of oxygen in effluent gas are reproduced in Fig. 6. This figure shows time courses of cell mass and sorbose concentrations, and partial pressure of oxygen in effluent gas. Batch growth of microorganisms occurred until an elapsed time of 18.5 hours. Thereafter, the feedback control was initiated as indicated by an arrow in Fig. 6. The sorbose concentration reached a constant value very quickly, while cell mass concentration took about 12 hours to reach the steady state value.

4.4.3 Carbon Dioxide Evolution Rate (CER) Controlled Bioreactors

Almost all microorganisms derive the energy required to convert the substrate into cell mass or product by oxidizing a portion of the supplied carbon source to CO_2. At steady state balanced growth conditions the energy producing pathways are coupled with the energy requiring pathways of cellular metabolism. There exisit a constant relationship between the substrate uptake rate and the CER. This idea is used in the CER-controlled bioreactors, in which the off gas CO_2 concentration (which is related to CER) is used to manipulate the nutrient feed rate to the bioreactor. Control based on CER was first demonstrated by Watson [8,20]. The operation was also found feasible for the continuous culture control of a methanol utilizing bacterium (*L3*) [21]. Experimental results of Lee [21] on the control of continuous culture of *L3* based on the CO_2 concentration in the effluent gas are presented in Fig. 7. As indicated by an arrow in Fig. 7, a PID control was initiated during a batch growth of *L3*. After a brief transient growth period of 30 hours the controller was able to stabilize the continuous culture as represented by the constant values of feedflow rate and cell mass and CO_2 concentrations.

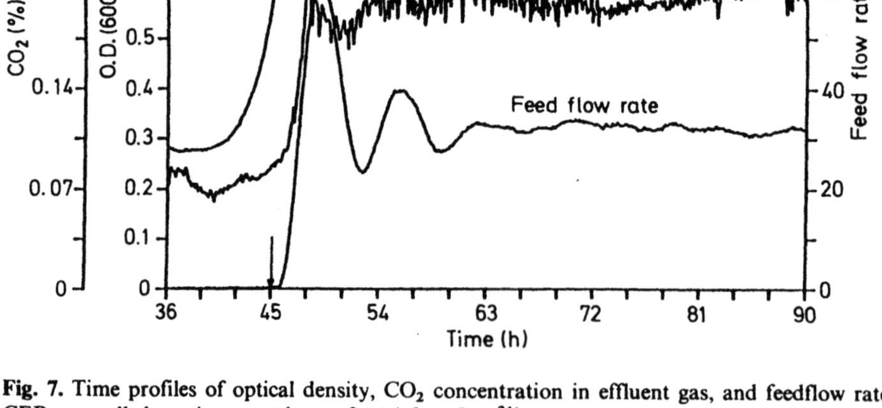

Fig. 7. Time profiles of optical density, CO_2 concentration in effluent gas, and feedflow rate in a CER controlled continuous culture of *L3* (after: Lee [21])

Bioreactor controls based on OAR, OUR and CER are similar to the BAR-controlled bioreactor. In each case the set value of the controlled variable determines the cell mass productivity, (μxv). General conclusions derived from the theoretical analysis of the BAR-controlled bioreactor also applay to OAR, OUR and CER controlled bioreactors.

5 A Novel Continuous Culture Operation

The turbidostat scheme for a conventional bioreactors is found to be unsuitable when the growth is substrate limited. This is not too surprising, because the control of cell density in a turbidostat is carried out by manipulating the flow rate of medium that contains the limiting substrate. In a turbidostat when the cell concentration is greater than the desired value, $x > x_d$, the controller would increase the dilution rate to increase the washout of cells, but this also increases the nutrient addition rate, Ds_F. Under the substrate limited growth condition, higher growth rates due to increased nutrient addition rates would slow the decrease in the cell concentration. Similarly, when $x < x_d$ the controller would decrase the dilution rate to slow down the washout of cells, but this also decreases the nutrient addition which would slow the increase in x. As a result of the two opposing effects, larger changes in the dilution rate are required to bring about smaller changes in the x value. This, obviously, is an undesirable characteristic of a controller. This drawback can be overcome in a continuous bioreactor with two feed streams. One feed stream contains the limiting substrate and the other stream contains all but the limiting nutrients. Alternatively, a pure water stream and a stream containing the concentrated amounts of the limiting substrate and other nutrients can be used. This novel continuous bioreactor operation is schematically shown in Fig. 8.

The cell mass and substrate balance equations for a bioreactor with two feed streams are

$$dX/dt = f_m(X, U) \tag{68}$$

$$X = \begin{bmatrix} x \\ s \end{bmatrix} \tag{69}$$

$$U = \begin{bmatrix} D_m \\ D_c \end{bmatrix} \tag{70}$$

and

$$f_m = \begin{bmatrix} -(D_m + D_c)\,x + \mu(s)\,x \\ D_C s_o - (D_m + D_c)\,s - \dfrac{\mu(s)\,x}{y(s)} \end{bmatrix} \tag{71}$$

At a non-trivial steady state ($x \neq 0$)

$$\mu = D_m + D_c = D \tag{72}$$

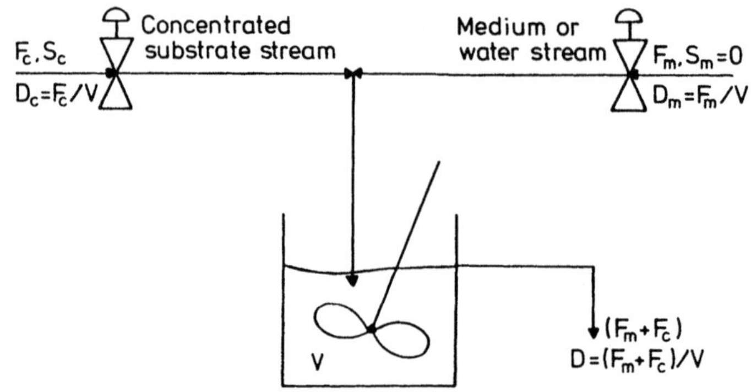

Fig. 8. A novel continuous culture operation

and the effective feed substrate concentration is

$$s_E = \frac{D_c s_c}{D_m + D_c} \tag{73}$$

and

$$x = y(s_E - s) = y\left(\frac{D_c s_c}{D_m + D_c} - s\right) \tag{74}$$

In this scheme both D_m and D_c can be controlled independently. Linearizing Eqs. (68) to (71) around a non-trivial steady state, we obtain

$$dX/dt = A_m X + BU \tag{75}$$

where

$$A_m = \partial f_m/\partial X|_{s.s.} \tag{76}$$

$A_m = A$ and is given by Eq. (11), and

$$B = \partial f_m/\partial U|_{s.s.} = \begin{bmatrix} -x & -x \\ -s & s_c - s \end{bmatrix}_{s.s.} \tag{77}$$

since det $B = -xs_c$ is non-zero, the rank of matrix B is two. Therefore, the rank of local controllability matrix for the system of Eq. (68) through (72), is also two. Hence, it can be concluded that all non-trival steady states are locally controllable. It should be noted that the local controllability criterion is met for all microbial systems whose steady state growth in continuous bioreactors can be modelled in terms of variables $\mu(s, x)$ and $y(s, x)$.

Consider the novel continuous bioreactor scheme in which the feed stream containing the limiting substrate is supplied at a constant rate, and the flow rate of other stream is manipulated to control the cell density. This scheme is called 'modified turbidostat' and is schematically shown in Fig. 9.

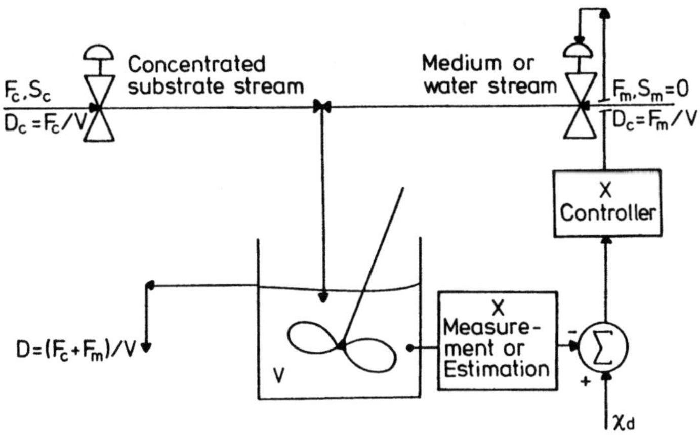

Fig. 9. A modified turbidostat control scheme

If a proportional control is used on D_m we have

$$D_m = D_{ms} + K_{cmx}(x_d - x) \geqq 0 \tag{78}$$

where K_{cmx} is the proportional gain. The steady state gain for the modified turbidostat operation is obtained from Eq. (74) as

$$K_{mx} = \frac{dx}{dD_m} = \frac{dx}{dD} = -\left[y\frac{D_c s_c}{\mu^2} + (y - y'x/y)/\mu'\right]_{s.s.} \tag{79}$$

Comparing Eqs. (31) and (79) we find that the steady-state gain of the modified turbidostat is higher than that of conventional turbidostat by yD_cs_c/μ^2. In the modified scheme the high steady state gains result at an expense of a lower saturation limit on the manipulated variable, D_m. For a constant μ the lower saturation limit on the manipulated variable for the conventional turbidostat is $D = \mu$, while for the modified turbidostat is $D_m = \mu - D_c$. In the modified scheme we choose D_c such that the product of the increase in the steady state gain yD_cs_c/μ^2, and the saturation limit on the manipulated variable, $D_m = \mu - D_c$, is maximized for a given μ. Therefore, D_c should be such that

$$\frac{\partial}{\partial D_c}\left\{(\mu - D_c)\, y\, \frac{D_c s_c}{\mu^2}\right\} = 0 \tag{80}$$

or

$$D_c = \mu/2 \tag{81}$$

As an example, for a microbial system described by model (3), to operate a fermentor at $\mu = 0.3\ h^{-1}$ and $s_E = 1.0\%$ w/v we choose $s_c = 2.0\%$ w/v, $D_c = 0.15\ h^{-1}$ and $x_d = 0.4\%$ w/v in the modified scheme. This situation is illustrated in Fig. 10 where the conventional turbidostat is compared with the modified turbidostat. In the region near $\mu = 0.3\ h^{-1}$ the conventional turbidostat would be ineffective because of very low steady-state gains, but the modified turbidostat scheme would be feasible since it has high steady state gains. For stable operation of the closed loop system K_{cmx} should be chosen positive.

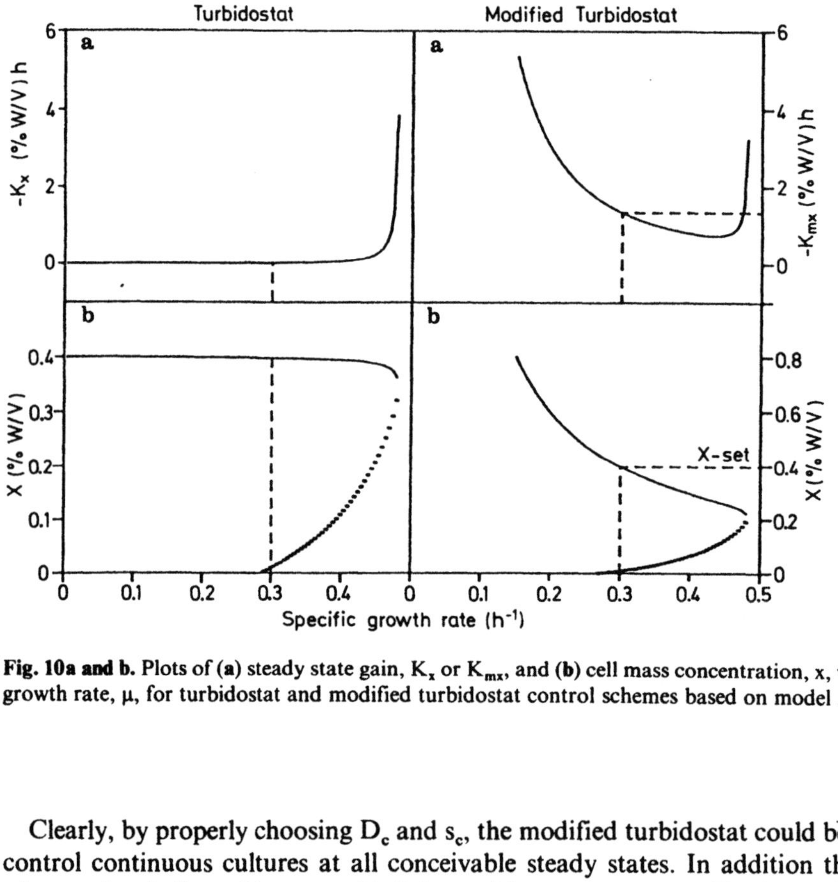

Fig. 10a and b. Plots of (**a**) steady state gain, K_x or K_{mx}, and (**b**) cell mass concentration, x, vs. specific growth rate, μ, for turbidostat and modified turbidostat control schemes based on model 3

Clearly, by properly choosing D_c and s_c, the modified turbidostat could be used to control continuous cultures at all conceivable steady states. In addition this mode of operation is free of input multiplicity and hence does not suffer from difficulties associated with it.

In the conventional bioreactor there are many variables that can be controlled (x, s, pH, BAR, OUR, OAR, and CER), but only one manipulated variable D. However in the novel continuous bioreactor operation proposed above, D_m as well as D_c can be manipulated. This possibility of multivariable control of bioreactors is currently being investigated by us.

6 Discussion

The conventional continuous bioreactor control schemes can be divided into two categories. Turbidostats, nutristats and pH-auxostats would comprise one category, in which the control schemes are feasible only at those conditions where growth is not substrate limited. Control schemes based on rates such as BAR, OUR, OAR and CER would constitute a second category. These control schemes are effective at almost entire steady state growth region, except around the maximum cell productivity for a microbial system that exhibits a constant cell mass yield. The rate controlled schemes also exhibit input multiplicity.

A novel continuous bioreactor operation is proposed. It is shown to satisfy the local controllability criterion at all steady states and for all microbial systems whose growth can be modeled in terms of variables $\mu(x, s)$ and $y(x, s)$. As a simple example of a control scheme for the novel bioreactor, a modified turbidostat is considered. As opposed to the conventional turbidostat, the modified turbidostat can operate under all growth conditions and is free of input multiplicity. Possibility of multivariable control of continuous bioreactors is also indicated. There are other control schemes that have not been analyzed here. These are, either restricted to a specific type of microbial growth as in a viscostat [54] for the production of xanthan bipolymers, or those that are still at a developmental stage, for example, a bioreactor control based on the redox potential [55]. There are also reports on the closed loop operation of bioreactors based on the respiratory quotient [56-57]. However, these have been applied to fedbatch or semibatch bioreactor operations only.

Acknowledgement: This work was supported in part by a grant from the National Science Foundation, CPE 7918261 and one of the authors (P.A.) was also supported in part by Purdue Research Foundation as a David Ross Fellow.

7 Nomenclature

A	system matrix or the Jacobian matrix [Eqs. (7) and (11)]
A_{III}	the Jacobian matrix [Eq. (76)]
a	constant in model-3
B	$m \times n$ matrix [Eqs. (7) and (77)]
BAR	base addition rate, mmoles per h
BAR_d	set point value of BAR
BC	buffering capacity, mmoles per l
b	$\partial f/\partial D$ [Eq. (12)]
b	constant in model (3)
C	$1 \times n$ matrix [Eq. (8)]
c^T	vector defined in Eq. (58)
CER	carbon dioxide evolution rate, mmole per l
c	constant in model (3)
D	dilution rate h^{-1}
D_s, D_{s1}, D_{s2}	steady state dilution rates, h^{-1}
D_c	F_c/v h^{-1}

D_m	F_m/v h^{-1}
DOC	dissolved oxygen concentration
F	total flow rate of medium, $l\ h^{-1}$
F_c	flow rate of concentrated stream, $l\ h^{-1}$
F_m	flow rate of medium or water stream, $l\ h^{-1}$
f	defined in Eq. (5)
f_m	defined in Eq. (71)
H^+	hydrogen-ion concentration, moles per h
H_d^+	set point value of H^+
K	proportional controller matrix [Eqs. (15)—(18)]
K_t^T	defined by Eq. (26)
K_n^T	defined by Eq. (39)
K_b	steady state gain of BAR-controlled bioreactor, mmoles
K_c, K_{cb}, K_{cn}, K_{ct}	proportional controller constants
K_m	constant in the Monod Model, $g\ l^{-1}$
K_{mx}	steady state gain of the modified turbidostat, (% w/v), h
K_s	steady state gain of a nutristat, (% w/v), h
K_x	steady state gain of a turbidostat, (% w/v), h
K_y	steady state gain
L_c	controllability matrix [Eqs. (9) and (13)]
L_c^0	output controllability matrix [Eqs. (10), (57) and (59)]
m	maintenance constant, h^{-1}
OAR	oxygen absorption rate, moles per h
OOC	off gas O_2 concentration
OCC	off gas CO_2 concentration
OUR	oxygen uptake rate, moles per h
(OH^-)	hydroxide-ion concentration, moles per l
r_{H^+}	rate of acid production per unit volume moles H^+ per h
r_{OH^-}	rate of base production per unit volume moles OH^- per h
s	substrate concentration, $g\ l^{-1}$ or % w/v
s_c	concentration of substrate in the stream F_c, $g\ l^{-1}$
s_d	setpoint value of S
s^E	effective feed substrate concentration, $g\ l^{-1}$
s_F	feed substrate concentration, $g\ l^{-1}$
t	time, h
U	vector of controls
v	volume, l
X	vector of states
x	cell mass concentration, $g\ l^{-1}$ or % w/v
x_d	setpoint value of x
Y	vector of outputs
y	cell mass yield, g cell per g substrate
y_0	yield constant in the maintenance model
$y_{H^+/x}$	yield of acid, moles H^+ per g cell mass
z	defined in Eq. (19)
z(0)	initial value of z
α, β	constants

μ	specific growth rate, h^{-1}
μ_m	maximum specific growth rate, h^{-1}
τ_i, τ_d	controller constant

8 References

1. Rolf, M. J., Lim, H. C.: Enzyme and Microbial Technology *4*, 370 (1982)
2. Dorby, D. D., Jost, J. L.: in: Annual Reports on Fermentation Processes (Perlman, D., ed.), Vol. 1, p. 95, New York: Academic Press 1977
3. Novick, A.: Ann. Rev. Microbial. *9*, 97 (1955)
4. Gerhardt, P.: J. Bact. *52*, 283 (1946)
5. Bryson, V.: Science *116*, 48 (1952)
6. Moss, F. J., Bush, F. E.: Biotech. Bioeng. *9*, 585 (1967)
7. Zines, D. O.: Biotech. Bioeng. *12*, 561 (1970)
8. Watson, T. G.: J. Appl. Chem. Biotechnol. *22*, 229 (1972)
9. Edwards, V. H., Ko, R. C., Balogh, S. A.: Biotech. Bioeng. *15*, 939 (1972)
10. Whaite, P., Gray, P. P.: Biotech. Bioeng. *19*, 575 (1977)
11. DiBiasio, D., Lim, H. C., Weigand, W. A.: AIChE J. *27*, 284 (1981)
12. Martin, G. A., Hempling, W. P.: Arch. Microbiol. *107*, 41 (1976)
13. Stouthamer, A. H., Bettenhaussen, C. W.: Arch. Microbiol. *111*, 21 (1976)
14. Driessen, F. M., Ubbels, J., Stadhouders, J.: Biotech. Bioeng. *19*, 821 (1977)
15. Oltmann, L. F., Schoenmaker, G. S., Reijndero, W. N. M., Stouthamer, A. M.: Biotech. Bioeng. *20*, 921 (1978)
16. McBean, R. D., Hall, R. J., Linklater, P. M.: Biotech. Bioeng. *21*, 1517 (1979)
17. Hospodka, J.: Biotech. Bioeng. *8*, 117 (1966)
18. Yano, T., Kobayashi, T., Shimiju, S.: J. Ferment. Technol. *56*, 416 (1978)
19. Yamada, S., Wada, M., Chibata, I.: J. Ferment. Technol. *57*, 210 (1979)
20. Watson, T. G.: J. Gen. Microbiol. *59*, 83 (1969)
21. Lee, C.: Ph. D. Thesis, Purdue University, W. Lafayette, IN 47907, USA 1981
22. Frederickson, A. G., Ramkrishna, D., Tsuchiya, H. M.: Mathematical Biosciences *1*, 327 (1967)
23. Fredrickson, A. G., Megee III, R. D., Tsuchiya, H. M.: Adv. Appl. Microbiol. *13*, 419 (1970)
24. Monod, J.: Ann. Rev. Microbiol. *3*, 371 (1949)
25. Andrews, J. F.: Biotech. Bioeng. *10*, 707 (1968)
26. Chen, B. J., Lim, H. C., Tsao, G. T.: Biotech. Bioeng. *18*, 1629 (1976)
27. Novick, A., Szilard, L.: Science *112*, 715 (1950)
28. Novick, A., Szilard, L.: Proc. Natl. Acad. Sci. *36*, 708 (1950)
29. Monod, J.: Ann. Inst. Pasteur *79*, 390 (1950)
30. Yano, T., Koga, S.: Biotech. Bioeng. *11*, 139 (1969)
31. Edwards, V. H.: Biotech. Bioeng. *12*, 679 (1970)
32. Spicer, C. C.: Biometrics *11*, 225 (1955)
33. Moser, M.: Proc. Natl. Acad. Sci. *113*, 222 (1956)
34. Koga, S., Humphrey, A.: Biotech. Bioeng. *9*, 375 (1967)
35. Herbert, D., Elsworth, R., Telling, R. C.: J. Gen. Microbiol. *14*, 601 (1956)
36. Mateles, R. I., Ryu, D. Y., Yasuda, T.: Nature *208*, 263 (1965)
37. Storer, F. F., Gaudy, A. F.: Enviro. Sci. Technol. *3*, 143 (1969)
38. Chi, C. T., Howell, J. A.: Biotech. Bioeng. *18*, 63 (1976)
39. Lee, E. B., Markus, L.: Foundations of Optimal Control, New York: Wiley 1967
40. Ray, W. H.: Advanced Process Control, New York: McGraw-Hill 1981
41. Koppel, L. B.: AIChE J. *28*, 935 (1982)
42. Agrawal, P., Lee, C., Lim, H. C., Ramkrishna, D.: Chemical Engineering Science *37*, 453 (1982)
43. Fuld, G. J., Dunn, C. G.: Ind. Eng. Chem. *49*, 1215 (1957)
44. Bach, H. P., Woehrer, W., Roehr, M.: Biotech. Bioeng. *20*, 797 (1978)
45. Neujahr, H. Y., Kjellen, K. G.: Biotech. Bioeng. *21*, 671 (1979)
46. Hikuma, M., Kubo, T., Yasuda, T., Karube, I., Suzuki, S.: Biotech. Bioeng. *21*, 184 (1979)

47. Kobayashi, T., Yano, T., Mori, H., Shimizu, S.: Biotech. Bioeng. Symp. *9*, 73 (1979)
48. Puhar, E., Guerra, L. M., Lorences, I., Fiechter, A.: Eur. of Appl. Microb. and Biotech. *9* (3), 227 (1980)
49. Hewetson, J. W., Jong, T. H., Grey, P. P.: Biotech. Bioeng. Symp. *9*, 125 (1979)
50. Luedeking, R., Piret, E. L.: J. Biochem. Microbial. Techn. Eng. *1*, 393 (1959)
51. Gaudy, Jr., A. F.: Biotech. Bioeng. *17*, 1051 (1975)
52. Brown, D. E., Halsted, D. J.: Biotech. Bioeng. *17*, 1199 (1975)
53. Hopkins, T. R.: Biotech. Bioeng. *23*, 2137 (1981)
54. Silman, R. W., Bagley, E. B.: Biotech. Bioeng. *21*, 173 (1979)
55. Kjaergaard, L., Joergensen, B. B.: Biotech. Bioeng. Symp. *9*, 85 (1979)
56. Wang, H. Y., Cooney, C. L., Wang, D. I. C.: Biotech. Bioeng. *27*, 975 (1979)
57. Wang, H. Y., Cooney, C. L., Wang, D. I. C.: Biotech. Bioeng. Symp. *13* (1979)

Modeling and Simulation of Bioreactor Process Dynamics

C. Kleinstreuer* and T. Poweigha
Department of Chemical and Environmental Engineering,
Rensselaer Polytechnic Institute, Troy, N.Y. 12181, U.S.A.

After a brief overview on biochemical reactors and associated processes, relevant model types and modeling approaches are discussed. In addition, a number of simple modeling applications focusing on bioreactor design, interfacial mass transfer and reactor control are reviewed.

In Part 2 published modeling and simulation work on biochemical reactors and reviews of related laboratory investigations are presented. Specifically, simulation case studies for gas-liquid reactors and hollow fiber bioreactors are discussed.

In Part 3, it is concluded that mathematical modeling and computer simulation are supportive tools for laboratory bench-scale investigations and are indispensable for the design, optimization, and control of large-scale bioreactors. However, existing models representing the process dynamics and economics of (research) bioreactors have to be greatly improved before the latter goal can be achieved.

1 System Description and Modeling Aspects

1.1 Introduction and Overview

In recent years, microbial conversion of cheap organic matter to valuable products other than foodstuff and alcoholic beverages has attracted a great deal of research attention. Although some activities in these areas are fueled by hopes for quick commercial success, there is a general motivation for basic and applied research in

* School of Engineering, NC State University, Raleigh, NC 27650

biotechnology. Specifically, in order to obtain laboratory-scale chemical production processes that suggest the feasibility of production of chemicals (using enzymatic catalysts) at the pilot-plant scale, the following research areas have to be stressed [105]:

- biocatalysis by genetically-improved, stable organisms;
- catalysis by biologically and chemically derived molecules;
- process and reactor vessel design; and
- product recovery, i.e. product separation and concentration.

Associated technical issues which have to be addressed, include:

- approaches for product release, separation and recovery, including strategies for reducing the water content of bioreactor streams for organics conversions;
- advances in microbiology and recombinant DNA techniques for genomic stability, resistance to harsh environmental conditions and enhanced productivity under reaction conditions;
- significance of feedstock selection on process performance;
- combining productivity/efficiency improvement from biological and chemical catalytic processes;
- new processes involving multiple species, either as separate organisms or as newly constructed, recombinant organisms;
- advanced concepts for control and regulation of biological processes;
- methods to extend the useful life of enzymes;
- control of secondary metabolite production and continuous processes; and
- process and reactor design and its relationship to productivity and energy efficiency.

To overcome the various technical barriers and to reach the overall goal, mathematical modeling and process simulation can be used as a cost-effective and reliable

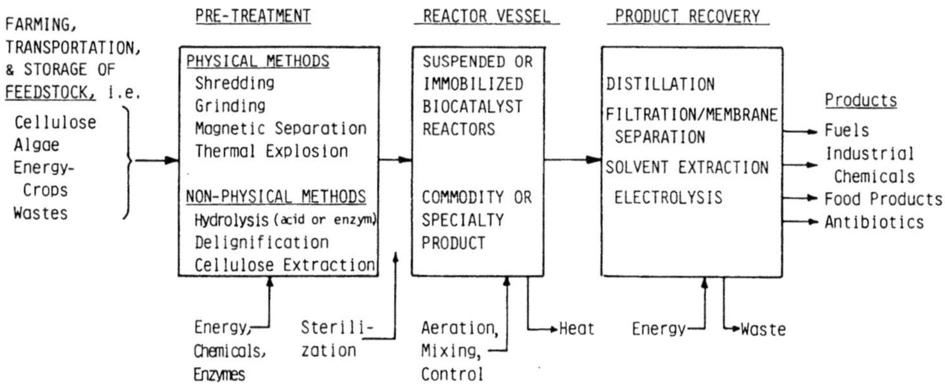

Fig. 1. Global biomass-to-product conversion system

tool at all stages of the biomass-to-product conversion cycle (Fig. 1). Specifically, computer simulation models aid in:

- detailed understanding of the process dynamics;
- prediction of system parameters and sensitivity analyses;
- evaluation of experimentally inaccessible states;
- calculation of upper and lower limits in: fluid flow, diffusion/mixing, and heat rejection;

- design of experiments and testing of hypotheses;
- assessment of capital investment and process costs; and
- process scale-up, optimization, and control.

Although the main emphasis in this review is on modeling and simulation of biochemical research reactors, a brief update is given on: bench-scale and industrial bioreactors with computer applications, mixing and scale-up, and product separation.

1.1.1 Biochemical Reactors

Traditionally, stirred tank batch reactors with controlled feeding of nutrients, roller bottles or continuous flow stirred tank reactor trains with recycle have been used in the bio-industry. More recent designs, still in the R & D state, include loop and air-lift fermenters, packed or fluidized bed reactors using biocatalyst beads, hollow fiber biochemical reactors and their spin-offs, and in an adaptation of a very old technique, solid-state fermenters [19, 74, 76, 93]. From an economical point of view, one has to distinguish if a process is cost-intensive in terms of feedstock, bioconversion, or product recovery. For example, intracellular proteins, antibiotics, vitamins and certain acids are products from *recovery* cost-intensive processes [28], i.e. the reactor economics such as capital and operating costs may play a minor role here. However, in cases where mass and heat transfer are the limiting steps, especially when high volumetric productivity is important, bioreactors play a central role in the overall biotechnological process. The basic goal of maximum production rate at minimum cost is usually difficult to achieve due to biochemical constraints (e.g. limited catalytic capacity of the cells/enzymes or limiting nutrients) and/or engineering constraints (e.g. upper bounds in momentum, mass and heat transfer). A few simple bioreactor design equations will illustrate this point.

Before bench-scale reactors are constructed and laboratory tests are conducted, it is important to reflect upon the objective function, operating conditions, and limiting step(s) of the given bioconversion system. For example, Cooney [28] and Wang et al. [111] among others discussed a number of useful formulas for continuous flow processes. They define the objective function to be maximized, the ratio of product per unit of biocatalyst, as an integral expression:

$$R_{p/x} = Y_{p/s} \int_0^{t_c} S_a(t)\, dt \tag{1}$$

where $Y_{p/s}$ is the conversion yield of raw material to final product which reflects selectivity, i.e. the performance of the biocatalyst(s); $S_a(t)$ is the specific activity of the cells or enzymes which together with t_c, the half-life of the catalyst, have to be sustained at high values. Obviously, clever bioreactor design and operation help to avoid catalyst inactivation, local substrate diffusion limitations, instabilities and high shear stress for microorganisms, so that high values for $R_{p/x}$ can be achieved. Typical numbers for $R_{p/x}$ range from 1 g of product per gram of catalyst for specialty products (e.g. drugs) to over 2500 for commodity products (e.g. corn syrup). Another handy formula for continuous, steady processes expresses the relationship between product concentration, P, and operating conditions such as the volumetric flow rate F:

$$P = \frac{S_a V X}{F} \tag{2}$$

where V is the reactor volume and X is the concentration of microorganisms. It can be easily seen that continuous flow reactors have to face the problem of low product concentration when a high volumetric throughput is required. For aerobic processes under steady-state conditions, the active cell concentration depends on the oxygen transfer which is the rate-limiting step:

$$X = \frac{Y_{O_2}}{\mu} k_L a(c^* - c) . \tag{3}$$

Biological constraints are given by Y_{O_2}, the oxygen yield for growth and product formation and by $\mu(S)$, the specific growth rate which is partially described by Michaelis-Menten kinetics. The rate of oxygen transfer is largely determined by k_L, the liquid film mass transfer coefficient and by a, the dispersion interfacial area per unit reactor volume. Finally, nutrient utilization and growth of microorganisms without inhibition are coupled via a non-linear first order rate equation which can be written for a transient process as:

$$\frac{dS}{dt} = - \frac{1}{\bar{Y}} \hat{\mu}(T) \frac{S}{K_s + S} X + \frac{F}{V} (S_0 - S) - \left(m + \frac{q_p}{Y_{p/s}}\right) X \tag{4}$$

where S is the concentration of substrate, $\bar{Y}$ is an average growth yield, $\hat{\mu}$ is the maximum specific growth rate, and K_s is the saturation constant, m is the maintenance requirement and q_p is the specific rate of product formation.

As mentioned earlier, several *research reactor* types have been reported in the literature, with applications in both research and industry. Differences between particular reactors often arise from the flow fields in the reactor and/or the mode of introducing the reactants. Two general classes of reactors are identifiable from the literature, namely *stirred tanks* and *tubular systems*. Further categorization of each class is possible on the basis of biocatalyst placement, mixing agent, and energy distribution system. For aerobic reaction systems, Tsao [104] listed several research reactor types. These include bubble columns, tower, tubular loop and microbial film fermenters, and stirred tanks. Some of these reactor types are also used in the chemical industry, e.g. bubble columns, tower fermenters, and the stirred tank reactor which is usually operated in the batch mode.

Recent trends in the development of bioreactors have involved the modification of tubular systems to incorporate advancements in biocatalysts (whole cells and/or enzymes), immobilization techniques, and membrane technology. Some of the new tubular biochemical reactors that exploit immobilized biocatalyst are the packed-bed (or fixed-bed), fluidized-bed, hollow-fiber, and membrane reactors. Several theoretical and experimental aspects of the operation of these reactor types have been presented in the literature in recent years, e.g. design criteria [74,82], determination of kinetic constants [63,106,109] analysis of electrostatic effects on kinetics [43,96], and calculation of effectiveness factors [115]. An alternative trend in the use of tubular systems for high-volume aerobic fermentation has been the development of gas-liquid reactors, exemplified by the airlift fermenter [45,70a,70b,71] and loop reactor [64,93].

The literature on *industrial reactors* is not as extensive as that on research reactors, probably for reasons of propriety. However, it is known that new designs are being developed, eventually to compete effectively with the traditional stirred tank batch-operated reactor (e.g. hollow fiber perfusion reactors, membrane reactors, multi-stage loop fermenters, and three-phase flow reactors). It is clear that continuous operation at high product concentration has to be adopted if the large production runs of the new biotechnology are to be achieved. Additional goals for industrial reactor design include high capacity, simple technology, reliability, and low energy expenditure.

It can be concluded from the literature survey that existing *bioreactor simulation models* are strongly based on empiricism and restrictive assumptions (Table 1). For example, rheological and non-isothermal effects are generally ignored and biokinetics and the temporal/spatial resolutions of the dependent variables (i.e. concentrations of substrate, nutrients, microorganisms and products) are often poorly represented. Momentum and mass transport phenomena are usually represented by empirical submodels which are only correct for a limited range of operational conditions. Reasons for these shortcomings include the lack of reliable and complete data sets as well as numerical solution constraints since the governing equations are usually available. Hence, ongoing modeling efforts in conjunction with laboratory experiments are concentrating on a better representation of the fluid mechanics (e.g. two- and three-phase flows, rheology of reaction mixture, etc.), transient mass transfer mechanisms and biochemical reactions in stirred reactor trains with recycle, airlift fermenters, hollow fiber biochemical reactors and fluidized as well as packed bed bioreactors.

In summary, desirable bioreactor performance and design criteria, most of which could be implemented in conjunction with advanced computer simulation models, include:

a) continuous high-volume feed with high product concentration;
b) high and fast conversion yield of feedstock to final product;
c) minimal by-product formation and environmental impact;
d) reactor flexibility to accommodate different cultures and modes of operation;
e) simple sterilization procedure and no danger of contamination;
f) low-level technology for minimum capital investment and maintenance;
g) on-line control for optimal operation, i.e. steady product quality and quantity;
h) permanent separation of microorganisms from product stream;
i) easy (built-in) separation and concentration of products; and
j) low-shear environment for sensitive microorganisms.

Some of the criteria are complementary, such as a) to c), and others are antagonistic, such as d), g) and h) versus f). For example, a given biosystem which performs very well in a specific reactor has usually lower yields in a different reactor. Also, not in all cases are continuous processes advantageous, mainly due to the problem of low product concentration. A decision tree for selecting biocatalysts and bioreactors with possible feed modes is given in Fig. 2.

1.1.2 Mixing and Scale-up

There is an extensive literature on mixing and scale-up, recent reviews of which have been presented by Nauman and Buffham [77] and Patterson [80] as well as by Moo-Young and Blanch [75], Schügerl [93] and Prokop [82]. Much of this knowledge however, applies to waterlike (Newtonian) fluids only. In biochemical engineering applications

Table 1. Computer simulation models of biochemical reactors

Reactor type	Process	Features	Selected references
CSTR	Anaerobic digestion	• One culture • Steady state • Isothermal • Liquid phase • Calculates yield coefficient	BIOTREAT [24] basically waste treatment oriented
CSTR	Anaerobic digestion	• One culture • Dynamic • Gas-liquid phases • Isothermal	1. Graef and Andrews [39] 2. Collins & Gilliland [27] 3. Buhr and Andrews [17]
CSTR	Anaerobic digestion	• Two cultures • Transient • Gas-liquid phases • Variable temperature	1. Kleinstreuer and Poweigha [61] 2. Cohen et al. [26]: Two reactors in series
Airlift reactor	Aerobic process (suspended microorganisms)	• Steady state • Isothermal • Axial variations only • No product kinetics	1. Hatch [45] 2. Merchuk et al. [71] 3. Deckwer [31]
Loop reactors Tower reactors Bubble columns	Aerobic process	• Transient • One-dimensional • Isothermal • Substrate & oxygen uptake, cell mass production	1. Luttmann et al. [64] 2. Schügerl [93] 3. Blenke [10] 4. Shah et al. [95]
Immobilized enzyme packed-bed	Glucose isomerization, with three common modes of catalyst immobilization: — adsorption on solid	• Effectiveness factors • Kinetics • Multiple steady states • Steady state • Transport parameters	1. Lee et al. [63] 2. Messing [72] 3. Klein and Vorlop [57]

Table 1. (continued)

Reactor type	Process	Features	Selected references
Immobilized enzyme fluidized bed	Particles (e.g. beads) — porous carrier, e.g. controlled pore glass — microencapsulation	• Tapered bed • Draft tube • Three-phase flow	1. Emery and Cardoso [35] 2. Scott and Hancher [94] 3. Hsu [51] 4. Andrews [3] 5. Fan et al. [37a, b]
Hollow fiber enzyme/whole cell reactor	Laminar axial convection with radial diffusion of substrate	• Effectiveness factors and Thiele Modulus • Linear reaction term • Substrate kinetics only	1. Rony [86] 2. Waterland et al. [113] 3. Kim and Cooney [55] 4. Webster and Shuler [114a] 5. Kleinstreuer et al. [60]

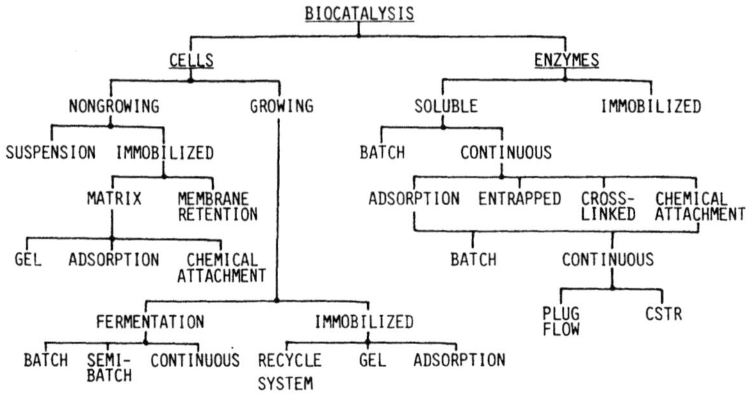

Fig. 2. Decision tree for selecting biocatalysts and bioreactor operation mode [28]

the liquid media are often highly non-Newtonian [100], so that the influence of rheology on mixing and scale-up have to be considered [9, 68, 73, 92, 111].

The major variables in scale-up studies are the mass and heat transfer coefficients, and the rheological parameters of the solution. The rheology pertains to flow characteristics, e.g. power requirements and two-phase flow parameters, and it plays an important role in the choice of product separation technique. Prokop [82] discusses the need for the development of a methodology of scale-up via modeling and simulation. He pointed out that similarity between *all* key parameters of the engineering *and* biological systems should be maintained.

Mixing in biochemical reactor systems is controlled by two factors which are unique to this area, namely, the relative fragility of the biocatalysts, and the rheology of the reaction broths. Turbulent mixing is usually desirable and achievable in low-viscosity fluid systems involving non-biological reactants/products. In biochemical reactor systems high shear rates could be detrimental to microorganisms and enzymes. High media viscosity may preclude the achievement of high levels of turbulent mixing (i.e. mass transfer) with reasonable levels of energy input.

In summary, mixing and scale-up are considered from the aspect of reactor type, fluid phases present, and fluid rheology, namely:

a) stirred tanks/anaerobic reaction systems;
b) stirred tanks/aerobic reaction systems — in this case mechanical and pneumatic agitation have to be considered separately;
c) tubular reactors/aerobic reaction systems;
d) tubular immobilized biocatalyst reactors/anaerobic reaction systems, and
e) non-Newtonian fluid systems.

In systems of type a) the main parameters for scale-up study are power requirements for mixing and the degree of non-ideal mixing [16, 13]. An additional key parameter for systems of type b) is the oxygen transfer rate via the overall mass transfer coefficient k_La.

In order to determine this parameter, gas hold-up and bubble dynamics also have to be considered; Calderbank [20] has provided extensive correlations for this class of reactors.

The class of tubular reactors with aerobic reaction systems is epitomized by the bubble column. Extensive studies have been undertaken but because of the complexity of the fluid dynamics of the system, only correlations are predominantly available, usually applicable to only narrow ranges of operational parameters. A variant of the tubular reactor which is rapidly gaining attention is the airlift reactor which is simply a bubble column with an internal baffle. Mathematical modeling, mixing, and scale-up aspects have been investigated for this type of reactor by Hatch[45], Merchuk et al.[71], Merchuk and Stein[70] and to a certain extent by Shah et al.[95]. By comparison, the extent of published work on mixing and scale-up of immobilized tubular biocatalyst reactors is far less than the two-phase aerobic systems. A probable reason is the ready availability of such studies in the traditional chemical reactor engineering literature of packed-bed reactors.

One factor that characterizes mixing and scale-up studies is the preponderant consideration of mass transfer to the virtual exclusion of heat transfer. One reason is that most studies and mathematical modeling efforts have only considered isothermal systems. Extensions to nonisothermal operation would be a logical avenue for further research.

1.1.3 Product Separation Techniques

Traditionally, the major products of interest in commercial applications of biotechnology have been enzymes, cell mass (e.g. baker's yeast, single-cell protein), and antibiotics. Consequently, satisfactory methods exist for the recovery of these products[1,6,111], especially from the point of view of the generally small production volumes. These methods include mechanical separation (settling, centrifugation, filtration, etc.), cell disruption to release intracellular products, and aqueous and solvent extraction, amongst the first-level or low-resolution separation techniques. Current high-resolution techniques include ultrafiltration, chromatography (gel and affinity), reverse osmosis, and electrodialysis (Table 2).

The anticipated large-scale production of biochemicals requires the improvement of current product recovery methods and the development of new ones to handle the large throughputs. Typical of the current techniques that require improvement is column chromatography. While the method is very highly selective and successful in relatively small columns, recovery efficiency drops dramatically in larger columns[18,79,116].

The current literature suggests that the product separation techniques of the future would be mainly the membrane processes, employed either alone or in combination with vacuum distillation processes[40,41,63]. The use of vacuum is to minimize the large energy requirement of traditional distillation operations. Without the use of vacuum techniques, the distillation step may render the production of some of the most attractive metabolites uneconomical[66]. Another avenue for an economic overall biomass-to-product conversion process is to integrate the product recovery step into the bioconversion step. From this point of view, the utilization of microporous membranes in hollow fiber biochemical reactors and membrane cell bioreactors have great potential.

A conceptual separation system using pressure-driven membrane units for product recovery is shown in Fig. 3.

Table 2. Typical product separation/recovery techniques

Technique	Typical products	Remarks	References
1. Mechanical methods • Cell disruption • Settling • Centrifugation • Filtration	• Enzymes • Antibiotics • Cell mass	• Well established • Slow (e.g. settling, filtration) • Potentially energy intensive (e.g. cell disruption	1. Aiba et al. [1] 2. Bailey and Ollis [6] 3. Wang et al. [111] 4. Rajagopalan & Tien [83]
2. Differential solubility and/or volatility • Distillation • Solvent extraction	Ethanol	• Well established technology • High energy costs with plain distillation • Use of vacuum to cut energy costs	1. Aiba et al. [1] 2. Gregor [40] 3. Gregor & Jeffries [41]
3. Chromatography • Affinity chromatography • Gel filtration chromatography • Size exclusion chromatography • Ion exchange chromatography	Enzymes	• High resolution in small units • Difficult to scale-up	1. Wang et al. [111] 2. Bailey and Ollis [6] 3. Yau et al. [116] 4. Ouano [79]
4. Membrane processes • Reverse osmosis • Ultrafiltration • Electrodialysis	Enzymes, Cell Mass, Antibiotics Process water recycle	• Wide scope for variations • Good potential for scale-up	1. Belfort [8] 2. Kleinstreuer and Belfort [58]
5. Others • Electrophoresis • Precipitation • Coagulation • Flocculation	Mostly enzymes	• Usually highly selective • Potentially large demand for materials, e.g. chemicals, flocculants, etc.	1. Wang et al. [111] 2. Bailey and Ollis [6]

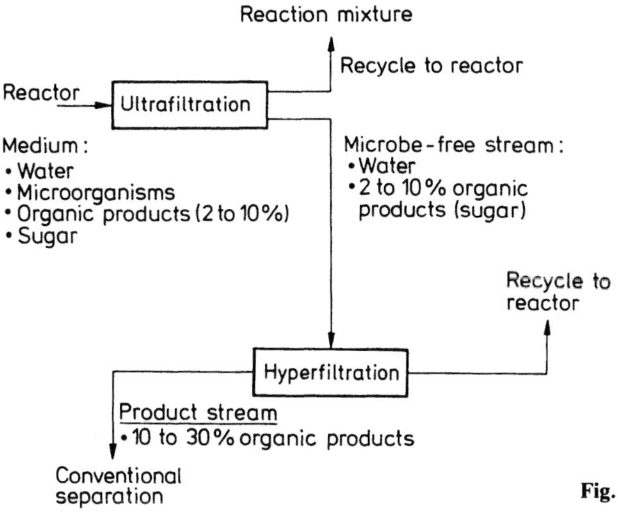

Fig. 3. Conceptual separator processes

1.2 Mathematical Modeling Aspects

As indicated in Sect. 1.1, simulation of the reactor process dynamics is imperative for the development of any bioconversion process where optimal economic and scientific results are expected. The desired degree of resolution for the computer simulation model and hence the theoretical effort required, depends on several factors. For example, financial and/or mathematical constraints often limit theoretical considerations for a *new* reactor design to a few hand-calculations using simple objective functions and semi-empirical relationships to estimate roughly the rate-limiting step, volumetric flow rate, glucose delivery rate, residence time, mixing intensity, interfacial mass transfer, pressure potential, and reactor volume (e.g. Prokop [82], Moo-Young and Blanch [76], Cooney [28], Schügerl [93], etc.).

In case a predictive mathematical tool is necessary, distributed parameter models which are comprehensive, flexible and accurate have to be developed. In this section, a brief overview on types of models and modeling approaches is given together with examples of computer applications in bioreactor control and data handling.

Models might be classified as verbal (concept, hypothesis, design, theory), physical (lab. bench-scale, analogue, pilot plant), or mathematical (deterministic, empirical, stochastic; continuous, discrete; analytical, numerical). As indicated in Fig. 4, mathematical models are, in general, either based on empirical data, deterministic principles, or stochastic phenomena.

Empirical models might serve as "order-of-magnitude" simulators of natural phenomena when only descriptive input/output-data sets are available and the system to be analyzed is too complex or too cost-intensive for deterministic or stochastic modeling approaches. In this sense, phenomenological relationships often surface as empirical submodels for coefficients and parameters in larger deterministic and/or stochastic models. The methodologies include: statistical/regression analysis, curve fitting techniques, mathematization of verbal concepts employing common sense and a mixture of all of them.

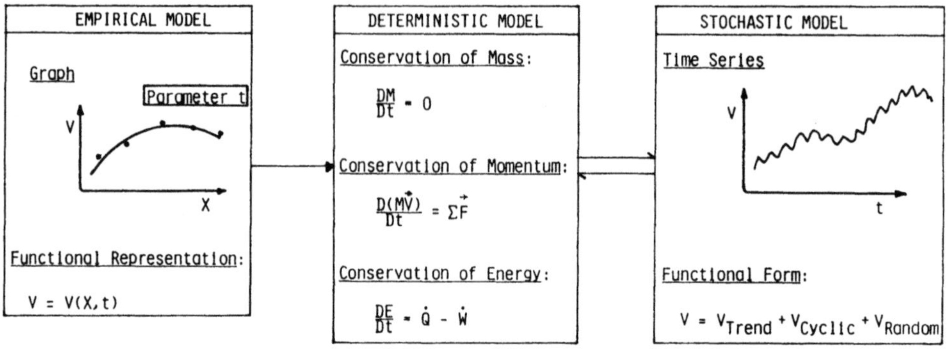

Fig. 4. Empirical, stochastic and deterministic modeling

A *deterministic model* is a direct mathematical representation of physicobiochemical phenomena which occur in deterministic, continuous or discrete patterns, i.e. the values of variables and parameters are definite fixed numbers. Approaches for the construction of deterministic models are either inductive, e.g. material or energy balances over a fixed control volume, leading to integro-differential equations or deductive, e.g. tailoring of generic partial differential equations.

Stochastic models are actually *the* tool for estimating the behavior of complex bio-engineering systems. Stochastic models which usually consist of random terms added to deterministic equations, admit elements of randomness in the mathematical description, so that the model results are probabilistic rather than fixed numbers.

The deterministic approach is almost exclusively employed in modeling biochemical reactors and is presented throughout this chapter. The *basic modeling steps*, illustrated with case studies in Part 2, are as follows (Fig. 5). The physical/biochemical mechanisms and unique problems of a given system have to be well understood. The system hierarchy, its boundaries and relations between all subsystems on equal and different hierarchical levels have to be recognized and mathematically specified. After the first step of problem definition and system identification is completed, governing equations and boundary conditions are postulated based on the system conceptualization (or analysis), project objectives, and mathematical and/or modeling constraints. Usually, the governing equations have to be (further) simplified due to the lack of suitable solution techniques or submodels to gain closure. After the final modeling equations are established, it is desirable to obtain two sets of reliable data either from laboratory/field studies or from analytical or asymptotic solutions of simplified systems. The first set provides input data for the model and profiles of dependent system variables with which the model can be calibrated. The second, independent data set is used to verify the model. However, in many cases, models are used without proper calibration and/or verification. On the other hand, models can be developed without experimental data to simulate phenomena so far only observed but not (yet) quantitatively described. The last modeling steps are the analytical or approximate/numerical solution of the tailored equations and the interpretation of the results. Finding the appropriate solution technique is often, next to parameter estimation, the most difficult problem although numerous equation solvers are available in computer libraries [108]. With the availability of computer graphics software, modeling results can be well presented in multi-color, and/or multi-dimensional form.

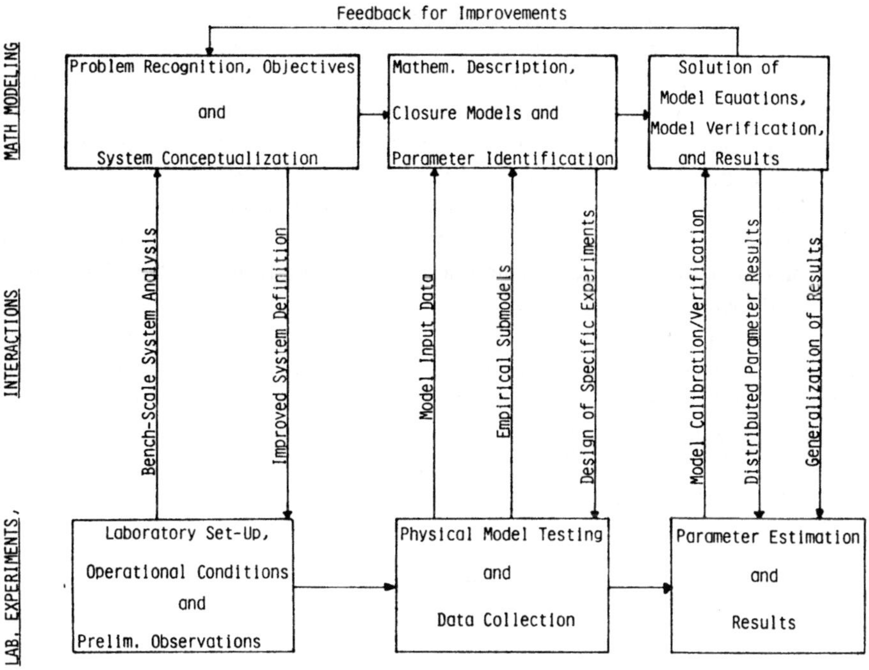

Fig. 5. Math modeling steps and interaction with laboratory investigations

Before specific (deterministic) bioreactor models are discussed in Part 2, fundamental approaches for systems analysis and the scientific tools for the development of modeling equations are reviewed.

There are two basic *system approaches* which are associated with Lagrange (closed system — constant mass) and Euler (open system — constant volume). The Eulerian point of view, natural to engineers, depicts a control volume fixed or moving in space for which a comprehensive force, mass and/or energy balance is taken. Now, systems classification for the open system approach (Euler) can be subdivided into (1) a lumped parameter, i.e. integral or macroscopical approach and (2) a distributed

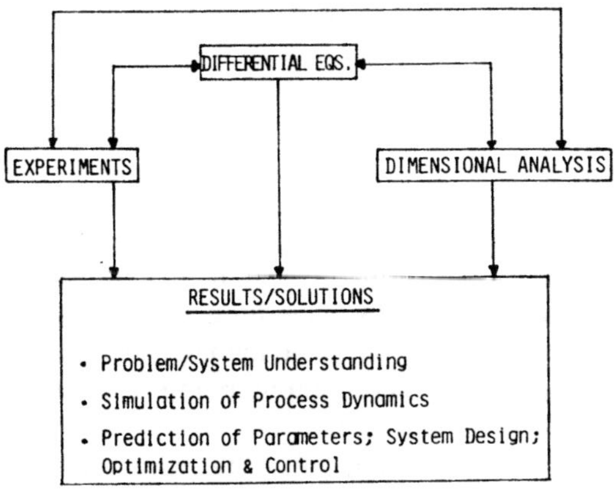

Fig. 6. Tools in engineering science research

parameter, i.e. differential or microscopical approach. The outcome of such a force, material or energy balance for a representative elementary volume of a continuous system, i.e. a small control volume which approaches (to) zero in the limit, are differential equations which represent the conservation of momentum, mass and energy. Differential equations together with experimental observations and dimensional analysis form the basic tools in theoretical systems analysis (Fig. 6).

Biotechnological systems are in general large and complex and hence they have to be decomposed and structuring principles for multifaceted system modeling should be applied. Such structuring principles include: (a) determination of systems hierarchy, subsystems and elements, i.e. decomposition of the real process; (b) determination of relations between levels of hierarchy, subsystems and elements; (c) development of models for all subsystems; and (d) integration of submodels into final mathematical formulation of the process dynamics, i.e. synthesis of all subsystems. Such a modeling methodology and applications to biomass conversion systems have been discussed by Zeigler et al. [117] and Prokop [81]. Figure 7 shows an example of several hierarchical levels of a biotechnological process as adapted from Prokop [81].

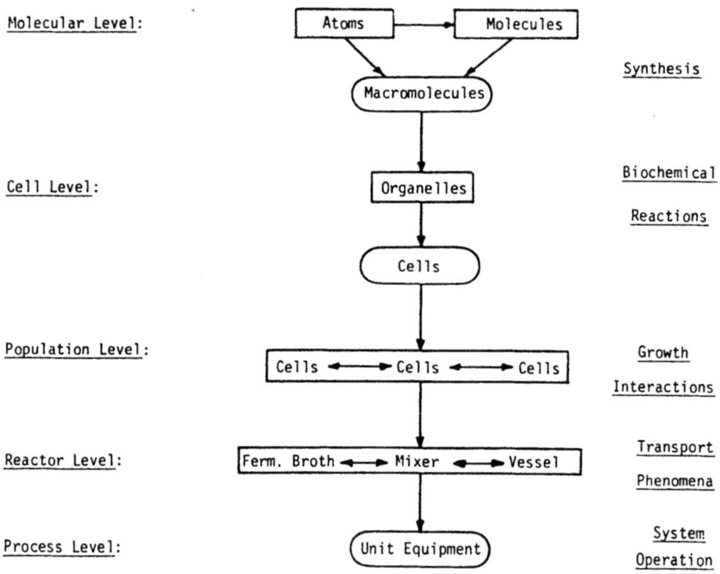

Fig. 7. Structural and function hierarchies of a bioconversion process (adapted from Prokop [81])

As indicated in Fig. 5, it is important that mathematical modeling interfaces with experimental investigations at various stages of the model development process. In the remainder of this section, the use of computers for data handling and bioreactor control are discussed. Figure 8 depicts the subtasks in data management [12, 107]. Hampel [44] gives an overview of the application of microcomputers interfaced with bench-top bioreactors for data reduction and analysis.

Given data sets are evaluated using a canned algorithm (i.e. a statistical package

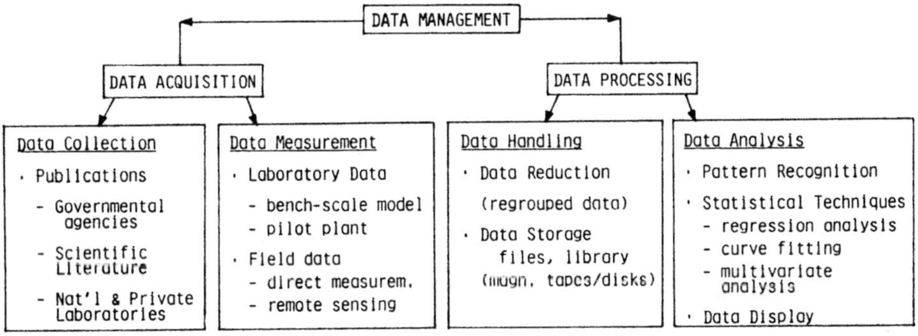

Fig. 8. Subtasks in data management (Vemuri [107])

for regression analysis) to produce the relative significance of all terms in a postulated equation usually an nth-order polynomial of the form: $y = \sum_{i=0}^{n} \alpha_i x_i^i$. The most important terms are retained and their coefficients are evaluated using the least-squares routine, for example. The resulting equation, an example of a lumped parameter, empirical model, possesses the best continuous fit of the discrete data sets when plotted. This empirical modeling approach is widely employed in parameter estimation and input/output-model design for complex and/or large systems. Hence, the procedure of establishing empirical models via curve fitting follows three basic steps: (1) postulating a functional form for the relationship between input and output variable; (2) estimating the number of significant terms, i.e. the degree of the polynomial in this example; and (3) determining the coefficients, here the α's, which yield an optimal fit through the data points.

A recent development in biotechnology is the application of computers in controlling the course of bioconversion processes, or monitoring biochemical experiments and estimating the reactor state from real-time measurements corrupted by white noise. In control applications the most common practice has been on-off control of physical variables of the biochemical process environment, e.g. temperature, flow rates, foam, etc. [1,111]. This has been feasible because quick-response sensors are generally available for monitoring these variables in real-time.

By comparison, computer control applications involving the use of feed-back loops based on measurements of the chemical environment are less widespread, although such methods are currently under study [5,11,56,82,93,99]. The reason for this is the general unavailability of quick-response sensors of the chemical environment of biochemical processes. Among the chemical variables that are currently measurable essentially in real time are: oxygen and carbon dioxide concentrations in the influent and effluent gas streams of aerobic biochemical reactors; dissolved oxygen concentration; and pH [111]. Sensors for other variables have been proposed recently, including alcohol [48], enzyme thermistor as an indication of biochemical reaction rate and glucose level [30]. As an example of on-line estimation of specific process parameters Stephanopoulos and San [99] presented a mathematical analysis of the state estimation problem and its application to the case of stirred-tank reactors. The underlying procedure in closed-loop control or on-line parameter estimation is:

a) measurement in real-time of some variables through which the state of the biochemical environment can be determined, b) postulation of some model of the system dynamics, and c) implementation of an optimization process to minimize an appropriate objective function. In parameter or state estimation, a typical tool is the Kalman-Bucy filter [14]. The status of computer control may be summarized as follows:

a) The availability of a larger number of quick-response sensors (in real-time) for the chemical/biological environment would greatly expand applicability [42].
b) Current state and parameter estimation procedures have been applied successfully to only non-product forming biochemical reactions and for stirred-tank reactors [99]; there is the need to extend the method to more general biosynthetic reactions and biochemical reactors; and
c) Computer control is a downstream operation in the sense that it requires the existence of an actual biochemical reactor with the computer and its associated data acquisition paraphernalia. It can only complement current off-line reactor modeling which is very much needed for design, scale-up studies, and the evaluation of process economics.

2 Computer Simulation Studies and Review of Experimental Data Sets

In Part 1 we discussed a number of modeling aspects in terms of types of mathematical models and modeling approaches in (biochemical) engineering. It was stated that deterministic models using a distributed parameter approach are favored for the simulation of the bioreactor process dynamics. For modeling purposes, one has to keep in mind that stirred tanks operated in *batch* mode or intermittent feeding or fed-batch mode are most frequently used in bioconversion processes; the exceptions include large single-cell protein plants and biotreatment of wastewater. *Continuous flow* stirred tanks with separation plus recycle and gas-liquid reactors have reached the pilot plant or industrial stage for high volume production of valuable biomass such as SCP. Immobilized or entrapped biocatalyst reactors including fluidized bed, packed bed, hollow fiber, membrane compartment, and microbial film bioreactors are still in the benchtop research stage. Herbert [47] concluded that continuous process for the production of biomass and primary metabolites are economically superior to conventional batch operations. Full acceptance of the continuous flow mode as well as new reactor designs hinge on the following requirements: (1) industry must be satisfied that process operation will be trouble-free; (2) the new processes must be economically superior to existing processes for making the same product; and (3) there must be a demand for products that existing processes cannot satisfy. In other words, researchers should recall that a new bioreactor design or operational mode has to be economically, rather than emotionally, attractive. It is evident that computer simulation models play an important role in achieving these objectives.

After a brief discussion of the basic modeling equations for bioreactor simulation, two case studies of biochemical research reactors with high potential will be reviewed.

2.1 Governing Equations for Simulation of Reactor Processes

As mentioned in Sect. 1.2, a force, material or energy balance over an infinitesimally small (reactor) volume, the *representative elementary volume* (REV), will lead to a set of coupled partial differential equations suitable for a distributed parameter approach. In contrast, a global balance over the entire reactor volume or its parts, leads to a set of integral equations which can usually be reduced to a system of (first-order) rate equations.

For the differential approach, the equations of motion and continuity, the species (or solute) mass transport equations, the energy equation, the equation of state including the biokinetic and species flux expressions are of interest. Larger "particles" (e.g. biocatalyst beads in fluidized beds) which do not follow the fluid motion are usually represented by monodispersed spheres. Hence, the field equations in terms of fluxes for a generalized reactor system can be written as

$$\text{(fluid continuity)} \quad \frac{\partial \varrho}{\partial t} + \nabla \varrho \vec{v} = 0 \tag{5}$$

$$\text{(motion)} \quad \frac{\partial}{\partial t} \varrho \vec{v} + \nabla \varrho \vec{v} \vec{v} = -\nabla p - \nabla \vec{\vec{\tau}} + \sum \vec{f} \tag{6}$$

$$\text{(mass transport)} \quad \frac{\partial c}{\partial t} + (\vec{v} \nabla) c = -\nabla \vec{j}_c \pm \sum S_c \tag{7}$$

$$\text{(internal energy)} \quad \frac{\partial}{\partial t} \varrho c_p T + (\vec{v} \nabla) \varrho c_p T = -\nabla \vec{q}_H - \vec{\vec{\pi}} : \nabla \vec{v} \tag{8}$$

The stress tensor $\vec{\vec{\tau}}$, the net sinks or sources S_o such as biochemical reactions, cell growth, oxygen transfer etc., as well as the mass and heat flux ($\vec{j}_c$ and $\vec{q}_H$) have to be specified for the system in terms of the principal variables $\vec{v}$, c and T in order to gain closure (see discussion below). An equation of state $\varrho = \varrho(p, c, T)$ is necessary when the assumption of incompressible flow does not hold. Suitable submodels for $\vec{\vec{\tau}} = \eta \cdot \vec{\vec{\gamma}}$ have to be found when the suspension flow exhibits non-Newtonian rheological properties. The energy dissipation term, $\vec{\vec{\pi}}$: $\nabla\vec{v}$ with the total stress tensor $\vec{\vec{\pi}} = p\vec{\vec{\delta}} + \vec{\vec{\tau}}$, couples the energy equation with the momentum equation explicitly.

In practical applications, Eq. (8) is usually ignored and semi-empirical expressions for single or two-phase flow velocity profiles are postulated in order to avoid solving (5) and (6). Hence, the principal modeling equation for simulating the reactor process dynamics, is Eq. (7) and its spin-offs. For example, transient convection-diffusion of species c_i in a multi-species two-phase flow field can be described by:

$$\text{(gas phase)} \quad \frac{\partial(\varphi c_i)}{\partial t} + (\vec{v}_G \nabla)\, c_i \varphi = D_{G,i} \nabla^2(\varphi c_i) + R_{G,i} + \sum_j I_{GL,j} \tag{9}$$

$$\text{(liquid phase)} \quad \frac{\partial}{\partial t}[(1-\varphi)\, c_i] + (\vec{v}_L \nabla)(1-\varphi)\, c_i = D_{L,i} \nabla^2[(1-\varphi)\, c_i] + R_{L,i} - \sum_j I_{LG,j} \tag{10}$$

where φ is the fraction of gas volume per unit volume of the system, R is the reaction term reflecting the biokinetics and I represents the interfacial mass transfer processes. Current research efforts concentrate heavily on realistic submodels for R and I (e.g. [82,93]).

In contrast to the differential balance where the REV → 0 and subsequently the field Eqs. (5) to (8) are produced, a *macroscopical balance* of forces, mass, energy and/or entropy for a finite control volume (e.g. reactor) generates an integral equation (Reynolds Transport Theorem):

$$\frac{D}{Dt} G_{syst} = \frac{\partial}{\partial t} \iiint_{c.v.} \eta \varrho \, dV + \iint_{c.s.} \eta \varrho \vec{v} \, d\vec{S} \tag{11}$$

$$\begin{Bmatrix}\text{Time rate of change}\\ \text{of arbitrary prop. G}\\ \text{as the syst. moves}\end{Bmatrix} = \begin{Bmatrix}\text{Accum. of specific}\\ \text{property } \eta \text{ within}\\ \text{control volume}\end{Bmatrix} + \begin{Bmatrix}\text{Net efflux of}\\ \text{specific prop. } \eta\\ \text{thru control surf.}\end{Bmatrix}$$

where $G_{system} = \iiint_{V_{syst}} \varrho\eta \, dV$ and η is equal to G per unit mass.

In addition to the necessary submodels for interfacial mass transfer and biochemical reactions, initial and boundary conditions have to be provided in order to gain closure for the modeling equations. Specific applications for gas-liquid, hollow fiber, and fluidized bed bioreactors will be discussed in the following sections.

2.2 Gas-Liquid Bioreactors

An advancement over sparged, mechanically agitated tank reactors are bubble columns and their spin-offs such as airlift, multi-stage tower or loop reactors, and three-phase fluidized bed bioreactors (Fig. 9). The advantages of such fermenters include better oxygen transfer and mixing, no mechanical agitation, and easier scale-up for certain configurations. The key parameters for gas-liquid reactor design and modeling are: the interfacial mass transfer coefficient, the gas holdup, the axial dispersion coefficients, the fluid fluxes, and the pressure requirement for sparging air into the reactor. Hence, the volumetric mass transfer coefficient, $k_L a$, and a power ratio (e.g. O_2-consumption rate versus power input) are often used to measure the reactor performance.

Shah et al. [95] reviewed empirical and semi-analytical expressions useful for the design of bubble columns generally employed in the chemical industry. Schügerl [93] reviewed the cultivation of yeast and bacteria in bench-scale tower loop reactors under different operational conditions. The influence of medium properties, construction parameters, and process variables on the oxygen transfer rate and cell mass productivity are considered. Hatch and Wang [46] discussed experimental and theoretical aspects of oxygen transfer in airlift fermenters. Merchuk et al. [71] presented a computer simulation model for the axial substrate, biomass and oxygen concentration profiles in an airlift bioreactor which is reviewed in the next Section.

2.2.1 Airlift Fermenter (Case Study)

Airlift bioreactors fall into the class of pneumatically agitated reactors or bubble columns. Airlift processes are more economical than stirred tank reactors for con-

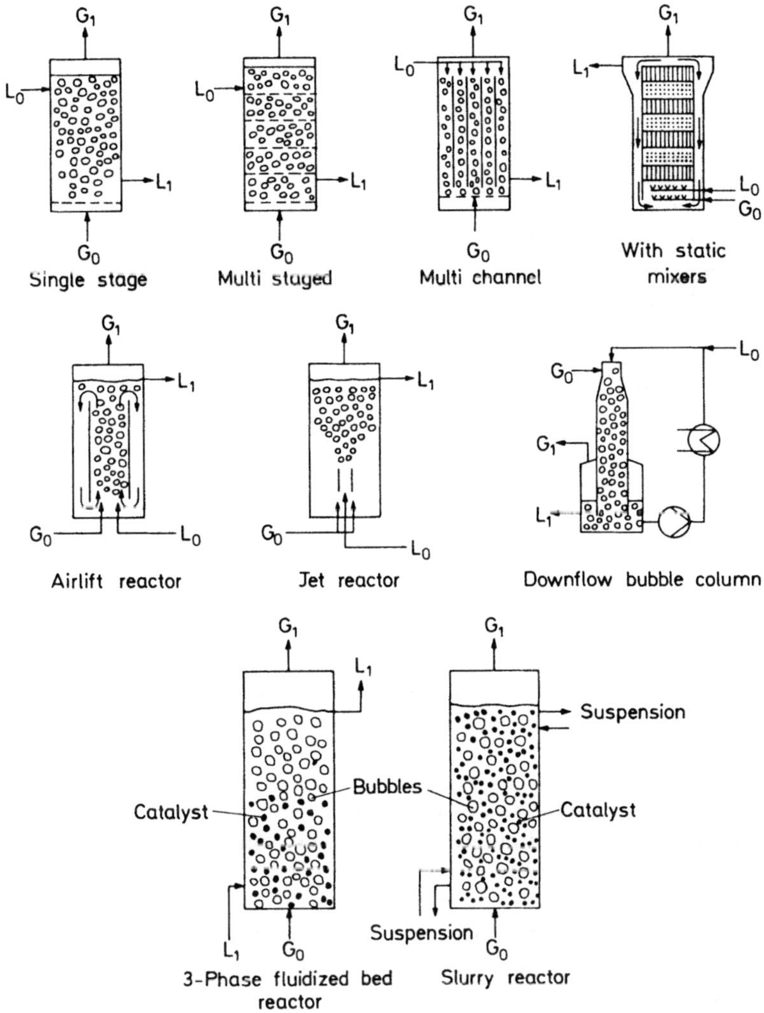

Fig. 9. Gas-liquid and three-phase bioreactors [95]

version processes with high oxygen demand such as SCP production. Sparging the filtered air into the solution sets up two-phase flow patterns where substrate, nutrients, microorganisms and products are well agitated and high oxygen transfer rates are achieved. Considerable backmixing in both phases, high pressure drop, bubble coalescence and reduced oxygen transfer into highly viscous fermentation broths can be disadvantageous in some instances [95].

This case study is based on the work by Merchuk et al. [71]. It illustrates in detail the modeling steps of system conceptualization, development of problem-oriented equations and closure, parameter determination, computer solution, and discussion of results.

The airlift fermenter consisting basically of two concentric cylinders and a centered sparger is shown in Fig. 10. It is assumed that two-phase flow occurs only in the

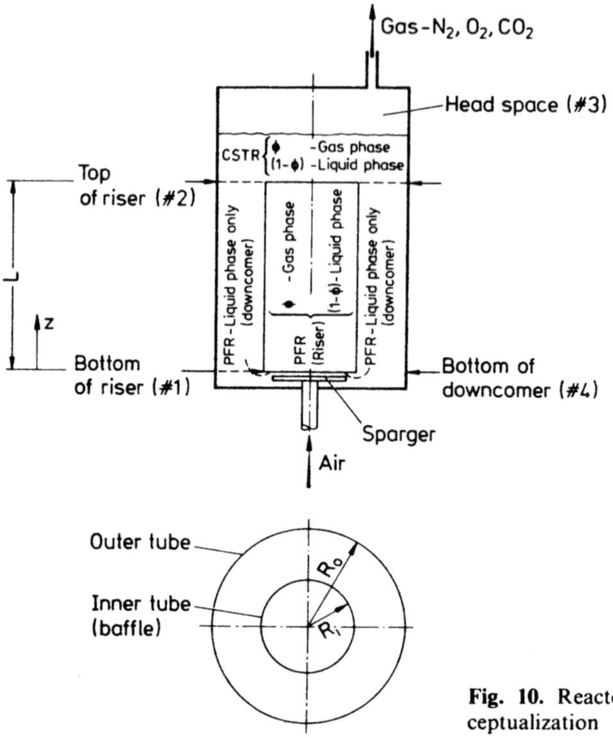

Fig. 10. Reactor geometry and flow field conceptualization

riser plus the head space which is regarded as a CSTR. Biological activities occur in all three regions of the reactor. The fluid dynamics are very simplified by postulating plug flow for both fluids (i.e. air and water) in the riser as well as in the downcomer, the annular liquid flow region.

The general equation describing the fermenter and other types of reactors were given in Sect. 2.1. Following the system conseptualization (Fig. 10), the governing equations for the *riser and downcomer* are:

$$\text{(gas phase)} \quad \frac{\partial(\varphi c_i^G)}{\partial t} + \bar{u}_G \frac{\partial(\varphi c_i^G)}{\partial z} = D_{G,i} \frac{\partial^2(\varphi c_i^G)}{\partial z^2} + r_{G,i} - \sum_k I_{GL,k} \quad (12)$$

$$\text{(liquid phase)} \quad \frac{\partial[(1-\varphi)\, c_i^L]}{\partial t} + \bar{u}_L \frac{\partial[(1-\varphi)\, c_i^L]}{\partial z} = D_{L,i} \frac{\partial^2[(1-\varphi)\, c_i^L]}{\partial z^2} + r_{L,i} - \sum_k I_{LG,k} \quad (13)$$

$$\text{(energy eq'n)} \quad \frac{\partial T}{\partial t} + (\vec{v}\nabla)\, T = k\, \nabla^2 T + \sum_i r_{L,i}(\Delta H)_{L,i} + \sum_k r_{L,k}(\Delta H)_{L,k} \quad (14)$$

where c is the concentration of substrate, biomass or oxygen, φ is the gas holdup, $\bar{u}$ is the averaged axial velocity, and T is the fluid temperature. The terms r and I represent homogeneous reactions and interfacial mass transfer, respectively. These equations describe the multi-species mass and energy transfer in two-phase plug flow reactors with axial dispersion.

For the *head space*, which is regarded as a continuously stirred tank reactor, the conservation principles can be stated as:

$$\text{(gas phase)} \quad \frac{d}{dt}(V_{SG}\varphi c_i^G) = F_{G,in}(c_{i,in}^G\varphi) - F_{G,out}(c_i^G\varphi) + V_{SG}r_{G,i} + \sum_k I_{LG,k} \tag{15}$$

$$\text{(liquid phase)} \quad \frac{d}{dt}[V_{SL}(1-\varphi)\,c_i^L] = F_{L,in}[c_{i,in}^L(1-\varphi)] - F_{L,out}[c_i^L(1-\varphi)] + V_{SL}r_{Li} - \sum_k I_{LG,k} \tag{16}$$

$$\text{(energy)} \quad \frac{d}{dt}[(V_{SG}+V_{SL})\,T] = F_{T,in}T_{in} - F_{T,out}T + V_{SG}\left[\sum_i r_{Gi}(\Delta H)_{Gi}\right] + V_{SL}\left[\sum_k r_{L,k}(\Delta H_{L,K})\right] \tag{17}$$

Equations (12) to (17) are equivalent to those presented by Cichy et al. [25] for two-phase tubular reactors and by Schaftlein and Russell [89] for two-phase stirred tank reactors.

In order to obtain a mathematically simple set of modeling equations, further assumptions, auxiliary equations for unknown parameters, and boundary conditions have to be postulated.

The simplifying assumptions incorporated by Merchuk et al. [71] are: steady state operation, constant gas holdup throughout the riser and head space, no axial dispersion, isothermal processes, and axial dependence only of principal variables.

The assumption of constant gas holdup, φ, deserves comment. A few researchers have reported both radial and axial variations of φ in bubble columns [95] which are similar to airlift reactors. Merchuk and Stein [70b] have proposed a procedure for measuring the axial variation of φ in the airlift reactor. Similar comments apply to the assumption about axial dependence only of principal variables, especially in large airlift reactors where it may be difficult to maintain plug flow profiles. Furthermore, it has to be noted that for aerobic reactions as considered in this study, the assumption of isothermal biokinetics may be unrealistic since such reactions are generally exothermic.

Once the global governing equations have been simplified, appropriate submodels have to be provided for the reaction and interfacial mass transfer terms in order to gain closure. We consider four system variables, namely, concentrations of substrate, biomass, and oxygen in the liquid phase, and concentration of oxygen in the gas phase.

The *microbial kinetics* are derived from:

(a) reaction stoichiometry:

$$\frac{1}{Y_s}[S] + \frac{1}{Y_{ox}}[O_2] = \text{Cell Mass (B)} + \text{Products (P)} \tag{18}$$

and
(b) substrate consumption and biomass growth kinetics:

$$r_B = \frac{\mu[B][S]}{K_m + [S]} \quad \text{(Monod Kinetics)} \tag{19 a}$$

$$\left.\begin{aligned} r_s &= -\frac{1}{Y_s} r_B \qquad &(19\text{ b})\\ r_{O_2} &= \frac{Y_s}{Y_{ox}M_{ox}} r_S \qquad &(19\text{ c}) \end{aligned}\right\} \text{(Consumption)}$$

Merchuk et al.[71] stated Eq. (19c) wrongly as:

$$r_{O_2} = \frac{Y_s}{M_s} r_s \,.$$

Product formation kinetics have not been considered. However, for primary metabolites, the product concentration profile would be proportional to the substrate concentration profile. Product formation in the case of secondary metabolites has to be considered separately.

For *interfacial mass transfer* only oxygen transfer is considered:
a) oxygen transfer rate [29]:

$$I_{LG,\,ox} = -k_L a \left(\frac{P_{O_2}}{H} - c\right); \tag{20}$$

this result is the standard two-film interfacial mass transfer model;
b) overall oxygen transfer rate [62]:

$$k_L a = 47.8 J_G^{0.997} \tag{21}$$

c) system hydrodynamics [45]:

$$\text{(gas volume flux)} \qquad J_G = \varphi V_G = \frac{Q_{gas}}{A_{riser}} \tag{22a}$$

$$\text{(liquid volume flux)} \qquad J_L = (1 - \varphi)\, V_L \tag{22b}$$

$$\text{(superficial gas velocity)} \quad V_G = 1.065 J + 32 \tag{22c}$$

$$\text{(superficial liq. velocity)} \quad V_L = 0.0152 J_G^{1.73} \tag{22d}$$

With the foregoing development of simplifying assumptions and phenomenological submodels the following *modeling equations* are derived (see Appendix AF for Nomenclature):
a) *Riser (PFR in both phases)* $z \in [0, L]$
Gas Phase:

$$(O_2\text{-pressure } p_{O_2}) \quad \frac{V_G\varphi}{RT}\frac{dp_{O_2}}{dz} + k_L a \left(\frac{p_{O_2}}{H} - c\right) = 0 \tag{23}$$

Liquid Phase:

$$(O_2\text{-conc. c})\quad V_L(1-\varphi)\frac{dc}{dz} + \frac{(1-\varphi)Y_s}{Y_{ox}M_{ox}}\frac{\mu BS}{K_m+S} - k_La\left(\frac{P_{O_2}}{H} - c\right) = 0 \tag{24}$$

$$(\text{substrate conc. S})\quad V_L\frac{dS}{dz} + \frac{\mu}{Y_s}\frac{BS}{K_m+S} = 0 \tag{25}$$

$$(\text{cell mass conc. B})\quad V_L\frac{dB}{dz} - \frac{\mu BS}{K_m+S} = 0 \tag{26}$$

The corresponding boundary conditions are:

$$at\ z = 0:\begin{cases} P_{O_2} = P_{O_{2,0}} \\ c = c_0 \\ S = S_a \\ B = B_0 \end{cases} \tag{27}$$

For the riser, z = 0 corresponds to the bottom of the riser and z = L to the top of the riser.

b) *Head Space (CSTR in both phases)*:
Gas Phase:

$$(O_2\text{-pressure})\quad V_GA_r\frac{\varphi}{RT}[P_{O_2,r} - P_{O_2,s}]$$
$$-\beta LA_r\left(1+\frac{1}{\xi}\right)k_La\left(\frac{P_{O_2,s}}{H} - c_s\right) = 0 \tag{28}$$

Liquid Phase:

$$(O_2\text{-conc.})\quad V_L(1-\varphi)A_rc_T - \frac{V_L(1-\varphi)}{\xi}A_rc_s$$
$$-\frac{Y_s}{Y_{ox}M_{ox}}\frac{B_sS_s}{K_s+S_s}\beta LA_r\left(1+\frac{1}{\xi}\right)$$
$$+\beta LA_r\left(1+\frac{1}{\xi}\right)k_La\left(\frac{P_{O_2,s}}{H} - c_s\right) = 0 \tag{29}$$

$$(\text{substrate conc.})\quad V_L(1-\varphi)A_rS_T - \frac{V_L(1-\varphi)}{\xi}A_rS_r$$
$$-\frac{\mu B_sS_s}{Y_s(K_m+S_s)}\beta LA_r\left(1+\frac{1}{\xi}\right) = 0 \tag{30}$$

$$\text{(cell mass conc.)} \quad V_L(1-\varphi)\,A_r B_T - \frac{V_L(1-\varphi)}{\xi} A_r B_s + \frac{\mu B_s S_s}{K_m + S_s}\,\beta L A_r\left(1+\frac{1}{\xi}\right) = 0 \tag{31}$$

c) *Downcomer (PFR, liquid phase only)*

$$\text{(O}_2\text{-conc.)} \quad \frac{V_L(1-\varphi)}{\xi}\frac{dc}{dz} + \frac{Y_s}{Y_{ox}M_{ox}}\frac{\mu BS}{K_m + S} = 0 \tag{32}$$

$$\text{(substrate conc.)} \quad \frac{V_L(1-\varphi)}{\xi}\frac{ds}{dz} + \frac{\mu}{Y_s}\frac{BS}{K_m + S} = 0 \tag{33}$$

$$\text{(cell mass conc.)} \quad \frac{V_L(1-\varphi)}{\xi}\frac{dB}{dz} - \frac{\mu BS}{K_m + S} = 0 \tag{34}$$

The corresponding boundary conditions are:

$$at\ z = 0: \begin{cases} c = c_s \\ S = S_s \\ B = B_s \end{cases} \tag{35}$$

For the downcomer, z = 0 corresponds to the head space and z = L to the bottom of the downcomer.

The modeling Eqs. (23) to (35) can be mathematically further simplified and written in a more compact form by implementing the following steps:

a) Calculation of the fluid dynamic parameters φ, J_L, V_L and V_G as a function of J_G by solving Eqs. (22a) to (22d) simultaneously.
b) Direct integration of the cell mass/substrate balance equations for riser and downcomer, i.e. Eqs. (23), (24) and (32) to (35).
c) Non-dimensionalization of the modeling equations using the results of (b) and defining:

$$S_A = \frac{S}{S_f}; \quad B_A = \frac{B}{Y_s \cdot S_f}; \quad Y_1 = \frac{P_{O_2}}{P_{O_{2,0}}}; \quad Y_2 = \frac{c_{O_2}}{c^*_{O_2}}; \quad \eta = \frac{z}{L} \tag{36}$$

and the following dimensionless groups:

$$\text{Stoichiometric Group} \quad D_1 = \frac{Y_s S_f H}{Y_{ox} M_{ox} P_{O_{2,0}}} \tag{37}$$

$$\text{Reaction Group} \quad R = \mu \frac{L}{J_L} \tag{38}$$

Stanton Number, Gas Phase $$St_G = \frac{Lk_LaRT}{J_GH} \quad (39)$$

Stanton Number, Liquid Phase $$St_L = \frac{Lk_La}{J_L} \quad (40)$$

Inserting the dimensionless variables and system parameters into Eqs. (32) to (35) yields the final modeling equations which are solved in a multistep procedure outlined in Fig. 11.

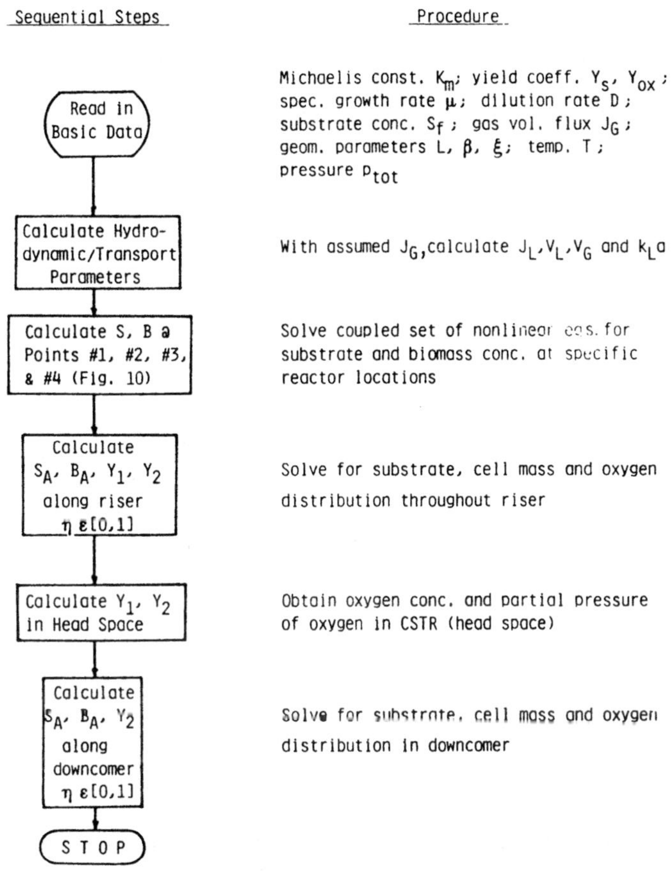

Fig. 11. Flow chart for solution procedure

Of the input data given in the Appendix AF, Merchuk et al. [71] did not state the values that they used for Y_s, β, and P_{tot}. Reference was made to a study by Goldberg et al. [38]) for "the stoichiometric constants". Merchuk et al. [71] did not use Y_{ox} since it did not appear in their derivations. For our model implementations we have used values for Y_{ox} from Wang et al. [111]. For β and P_{tot} we used values around 0.1 and 1.5 (atm) respectively.

In generating model results for sensitivity analyses, Merchuk et al. [71] used the reaction group $R = \mu L/J_L$ as a parameter. It is clear that R depends on the independent variables J_G and L, for a given microbial species i.e. a fixed μ. Different combinations of L and J_G may give the same value of R but the J_G values may correspond to different flow regimes within the reactor. Alternatively, the values of L may not correspond to consistent reactor sizes. It is thus useful to ensure that the combinations of the defining variables for R are consistent.

The phenomenological submodels that have been used in deriving the model equations include microbial kinetics, two-phase flow hydrodynamics, and interfacial mass transfer. All the submodels are steady-state and/or lumped parameter models, and so should not apply to other than steady-state system models. As it is well-known, non-steady state situations are the rule rather than the exception in most practical cases. This fact immediately exposes one of the major limitations of the model.

The results of the model implementation are summarized in three graphs, Figs. 12–14. Key input data for these runs are presented in Appendix AF. Figure 12 presents

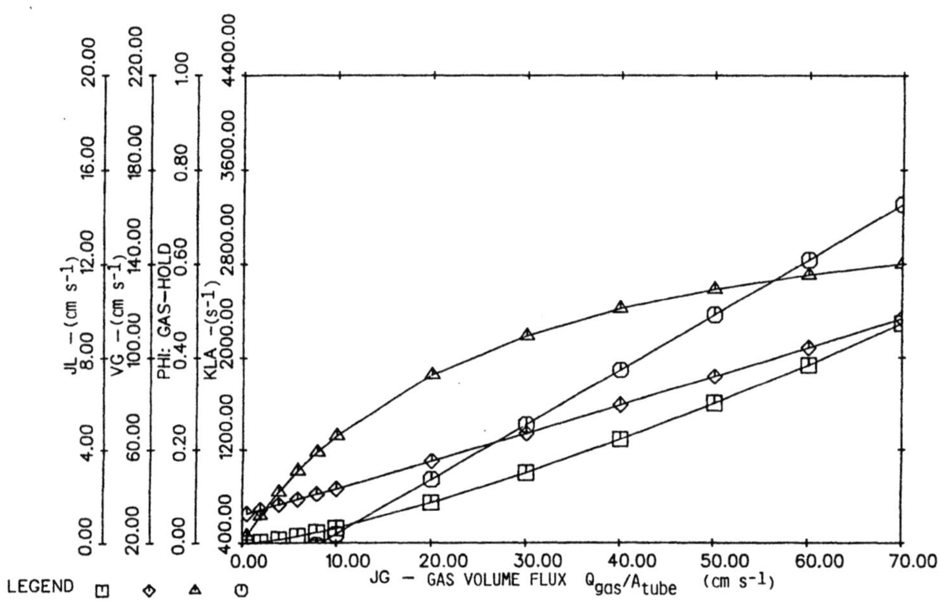

Fig. 12. Fluid dynamic and mass transfer parameters for conventional airlift reactor [71]

the variation of key average fluid dynamic/transport parameters within the reactor, as a function of the gas volume flux, J_G. It is seen that J_L, V_G, and k_La all change almost directly with J_G; the gas hold-up, φ, on the other hand, appears to taper off with J_G. As was discussed earlier, the hydrodynamic submodel employed is quite idealized; one of the reasons is the difficulty of adequately describing the two-phase flow hydrodynamics within the reactor. It is expected that further increases in J_G would cause a discontinuity in the value of each parameter as the flow field changes into a different regime, e.g. bubble flow into slug flow.

Figure 13 shows the axial variation of the nondimensionalized system state variables, substrate, S_A, biomass, B_A, gas phase oxygen partial pressure, Y_1, and dissolved oxygen, Y_2. The region 0 to 1.0 for η corresponds to the riser section of the reactor;

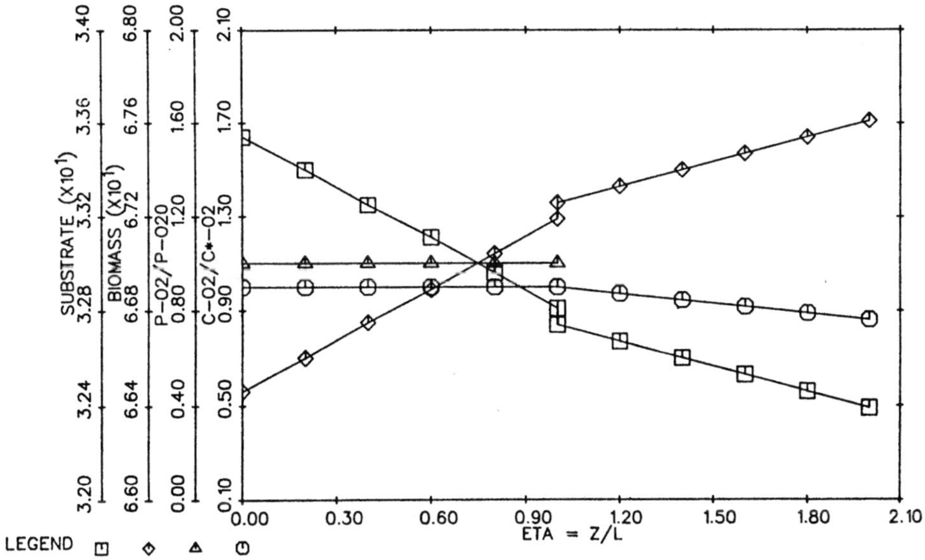

Fig. 13. Axial variation of system state variables in airlift reactor

the region 1.0 to 2.0 corresponds to the downcomer. The variation of each variable is consistent with physical expectations. For example, the substrate level decreases progressively from the bottom of the riser to the bottom of the downcomer because of its consumption in microbial growth. The discontinuity at $\eta = 1.0$ in each of the variables represents the conceptualization of the head space region as a CSTR; we have in effect, a PFR-CSTR-PFR train corresponding to the riser-head space-downcomer sequence. It has to be noted that both Y_1, and Y_2 change only slightly. In the riser section both variables are virtually unchanged. This represents a low metabolic rate of the microorganisms with a correspondingly low oxygen demand. It is assumed that all free oxygen is expelled from the head space region; thus only the Y_2 profile is indicated in the downcomer region. In this region, the higher biomass concentration offsets the reduced substrate concentration to cause a higher demand for oxygen, as indicated by the perceptible change in Y_2. A different set of basic input parameter would, of course, change the actual variations in each of the state variables although the qualitative trends would remain the same.

Figure 14 shows the variation of the substrate, S_A, and biomass, B_A at the bottom of the downcomer with dilution rate D. Also indicated are the products S_A*D and B_A*D, which may represent the substrate consumption and biomass productivity respectively. This figure corresponds to Fig. 5 or Merchuk et al. [71]. The trends in S_A and B_A agree with physical observation. The dilution rate is the inverse of the residence time so that as D increases, there should be a correspondingly lower conversion of the substrate, and lower biomass productivity. This is represented by high levels of unconsumed substrate and low levels of effluent biomass respectively, as in Fig. 14.

A final comment about the results pertains to the parameter ranges. As should be expected, each parameter must lie within an appropriate range of values if a solution of the model equations is to be guaranteed. Typical indicators of inconsistent

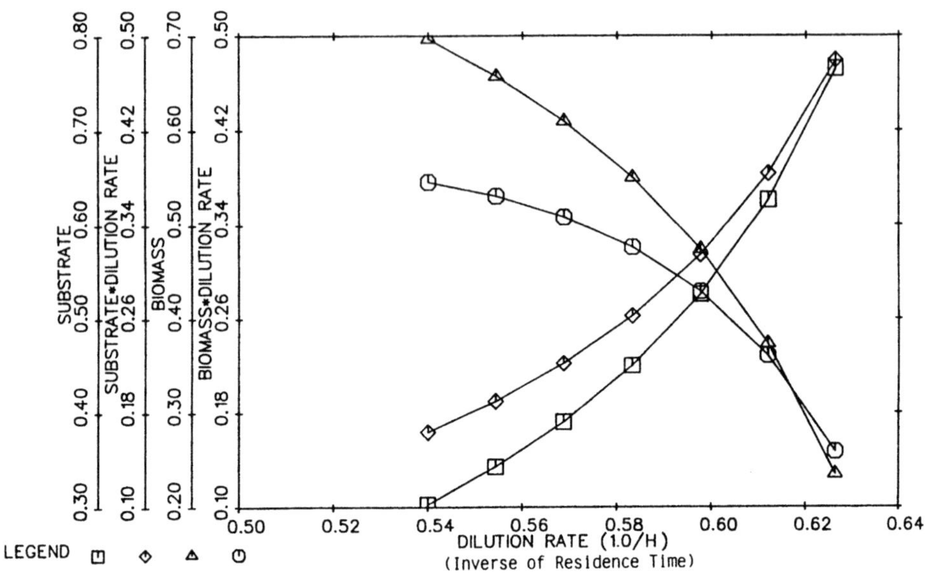

Fig. 14. Substrate and biomass at bottom of downcomer

parameter values include the occurrence of negative variables for logarithmic function evaluations, and the failure of the nonlinear equation solutions to converge. Examples encountered include: a) a maximum value of 775 mg l^{-1} for the substrate feed concentration, b) a maximum value of 4 for ξ, the ratio of cross sections of riser to downcomer; and c) a maximum value of 1000 cm for the height of the riser. These terminal values correspond to the particular choices of the other variables. It is reasonable to expect that other choices would give a different set of terminal values. The problem of determining the boundaries of the acceptable parameter-space, a priori, if possible, is quite appealing; however, in models where the parameters are highly coupled with the system state variables, as we have here, a systematic search technique would be necessary to keep computer cost reasonable.

Some of the possible avenues for the extension of the model presented in this case study include:

a) The development of a non-steady state submodel for microbial kinetics covering the entire range of the mircobial growth cycle.

b) The consideration of the interfacial mass transfer processes from the viewpoint of irreversible thermodynamics which appropriately applies to the non-steady state situation. The objective would be to obtain a submodel for k_La, which incorporates system transients, e.g. hydrodynamics, temperature, etc.

c) A hydrodynamic submodel for the two-phase flow which can describe both radial and axial variations in the flow parameters, e.g. gas hold-up, φ; in addition, rheological properties of non-Newtonian solutions should be incorporated.

d) Non-isothermal operation. The goal would be to account for the temperature dependence of the microbial kinetics and possibly of the mass transfer parameters. Since aerobic processes are in general exothermic, it is reasonable to expect that significant temperature changes would arise to negate the assumption of isothermal operation.

e) Inclusion of carbon dioxide transport. This would be useful since significant levels of the production of this gas accompanies the microbial growth, and the gas bears an important relationship to the ionic state of the liquid environment. Such an inclusion would enable the calculation of pH effects on the microbial kinetics.
f) The use of reactor flow submodels that are intermediate between the extreme cases of complete mixing and plug flow. The material is essentially available and only requires incorporation into the overall model of the airlift reactor.

2.2.2 Experimental Data

Although some progress has been made in the modeling of airlift reactors, experimental data on its operation, especially on a large scale, have been sparse by comparison. For the most part, experimental investigations concentrated on mass transfer and hydrodynamic aspects and reactors with internal baffles and external loops. Hardly any data sets are available on the heat transfer aspects although the usually exothermic aerobic reactions could pose a heat dissipation problem.

Hatch[45] measured the oxygen transfer in an airlift reactor with internal baffle using a culture of *Candida intermedia*, sub-strain of NRRL Y-6328-1. Hydrodynamic and transport parameters were determined for each region of the reactor, the draft tube (or riser), head space, and annulus (or downcomer). Typical of his results are the correlation equations for the volumetric mass transfer coefficient and gas holdup in the riser. These correlations have been used in Sect. 2.2.1.

Following the work of Hatch[45], a few research groups have emerged, amongst them Deckwer and Schügerl and their associates. Substantial work has been done by several other research teams as well; Table 3 summarizes the results.

The table indicates four data sets for the gas hold-up, φ, besides the work of Hatch[45]. Merchuk and Stein[70] measured φ and observed axial variations in this quantity; they developed a set of equations from which φ can be calculated as a function of hydrodynamic parameters. These equations correspond to those used to calculate φ in the series of airlift reactor models published by Merchuk et al.[70,71]. However, these data do not correspond directly to the simulation results since the system of equations contain a system-specific constant b, and the size of the experimental system was different from the simulated system. Deckwer et al.[33] measured the gas holdup as well as the volumetric mass transfer coefficient, and the liquid phase dispersion coefficient, in CMC solutions. Their goal was to obtain data under liquid rheological properties that are similar to those in actual fermentation broths. Hills[49] measured the hold-up for fluid throughputs that were very high compared to most other studies, but which better represent the actual flow rates in airlift reactor operations.

Volumetric mass transfer coefficients have been measured by several groups as indicated in Table 3. Orazem and Erickson[78] did not propose any correlations but they included in their paper comparisons of oxygen mass transfer efficiency and performance ratio for several aerobic reactor systems. These comparisons are presented as Figs. 15a and b. Margaritis and Sheppard[67] used single- and double-draft tube airlift reactors for their measurements. The results presented apply to the single-tube case for comparison with the other data which were all obtained with a similar type of reactor configuration. The dimensions of their system, height-to-diameter ratio of 1.5, are very different from those reported for most other systems. The research

Table 3. Summary of experimental data and empirical submodels for airlift reactors (Symbols are defined in Sect. 2.2.1)

Parameter/references	Equations/results	Remarks
1. Gas holdup φ	$V_L = \frac{J_L}{1-\varphi} = bJ_G^n$	• To calculate φ, solve the system of equations
Merchuk and Stein [70b]	$V_G = \frac{J_G}{\varphi} = C_0J_M + V_D$	• Authors observed an axial variation of φ
	$J_M = J_L + J_G$	• Height of reactor 4.05 m, H/D_i — ratio $\approx$ 15
	n Constant for flow regime ($\sim$0.4) b Constant for reactor (13–34) $C_0 \approx 1$; $V_D \approx 0.33\ m\ s^{-1}$	• Internal cylindrical baffle
Deckwer et al. [33]	$\varphi = 0.0265U_G^{0.82}$, $U_G \geqq 2\ cm\ s^{-1}$	• For CMC solutions • Reactor height 270 cm, $H/D \approx 20$
Hatch [45]	$\varphi = j_G/(C_0 + V_{rd})$ $j_G = Q_g/A_d$ C_0 = Constant ≈ 1.0 $V_{rd} = U_G - U_L$	• Result for draft tube • Height of reactor 500 cm, $H/D \approx 20$ • For media using *Candida intermedia*
Erickson and Deshpande [36]	• Presented results showing φ variation with axial position • No equation or correlation	• Used CMC solutions to test viscosity effects • Split cylinder baffle • Tested effects of surfactants also
Hills [49]	$\frac{U_G}{\varphi} = 0.24 + 1.35\ V_m^{0.93}$ when $U_L > 0.3\ m\ s^{-1}$ $U_G - U_L = 0.24 + 4.0\varphi^{1.72}$ when $U_L \leqq 0.3\ m\ s^{-1}$ $V_m = U_G + U_L$	• Used external loop • High throughputs, U_G from 0.07–3.5 $m\ s^{-1}$, U_L from 0–2.7 $m\ s^{-1}$

2. *Volumetric mass transfer Coefficient* k_La Orazem and Erickson [78]	Did not propose any correlations but presented a graph comparing this study to others	Figures 2.7a and 2.7b are useful: oxygen transfer coefficient, and performance ratio, respectively
Deckwer et al. [33]	$k_La \doteq 2.08 \times 10^{-4}\, \bar{u}_G^{0.54} \bar{\eta}^{-0.84}$ $\bar{\eta} = K(50\bar{u}_G)^{n-1}$ $\dot{\gamma} = 50\,U_G$	• Results for CMC solutions • Reactor height 270 cm, H/D ~ 20 • Tested k_La in fermentation broth with *Penicillium chrysogenum* • Derived result from O_2 conc. profiles
Margaritis and Sheppard [67]	• $k_La = b\left(\frac{P}{V}\right)^n$ (h^{-1}) $b \approx 0.200 - \sim 4.0 \times 10^{-2}$; $n \approx 1.0$ • $\left(\frac{P}{V}\right)$ = HP input per unit volume • $P = \frac{0.0643\,T_1 Q\gamma}{520(\gamma - 1)}\left[\left(\frac{P_2}{P_1}\right)^{\frac{\gamma-1}{\gamma}} - 1\right]$ • $\gamma = C_p/C_v = 1.395$ • Power economy, E: $E = AV_s + B$, $A \cong 0.012$, $B \approx 0.9–1.5$	• Used single and double draft tubes. Indicated results are for single tube • Height of liquid column 0.285 (m) H/D ratio 1.5 • Used sodium sulfite as O_2 sink; i.e. no bacteria
Hatch [45]	$k_La = \frac{1.39 \times 10^4 (j_g)^{0.57} (Qu)^{2.72}}{[V_r(T_m)^{0.1}}$ $T_m = \varphi_d H_g / j_{gd}$, Residence time of gas in draft tube $Qu = j_g/(j_g + j_L)$ $V_r = U_G - U_L$	• Same system as listed under gas hold-up

Table 3. (continued)

Parameters/references	Equations/results	Remarks
Luttmann et al. [65]	$k_La(x,t) = \begin{cases} K_La(t)\exp(-K(t)\,x/\alpha L), & 0 \leqq x \leqq \alpha L_R \\ K_La(t)\exp(-K(t)), & \alpha \cdot L_R \leqq x \leqq L_R \end{cases}$ $K(t)$ = Coalescence factor $k_La(t)$ = Value of k_La at reactor entrance	• Results include axial profiles of substrate, oxygen, and biomass concentrations
3. *Dispersion coefficient* D_L Deckwer et al. [33]	$D_L = 2.7 d_c^{1.4} U_G^{0.3}$	• For CMC solution • Reactor height 270 cm, H/D ≈ 20
4. *Heat transfer numbers* Deckwer [32]	$St = 0.1(Re * Fr * Pr^2)^{-0.25}$ $St = h/\varrho C_p U_G$, Stanton number $Re = U_G d_b/\nu$, Reynolds number $Fr = U_G^2/g d_b$, Froude number $Pr = \nu/a$, Prandtl number $a = k/\varrho C_p$, thermal diffusion coeff.	• Tested systems other than aerobic bioreactors

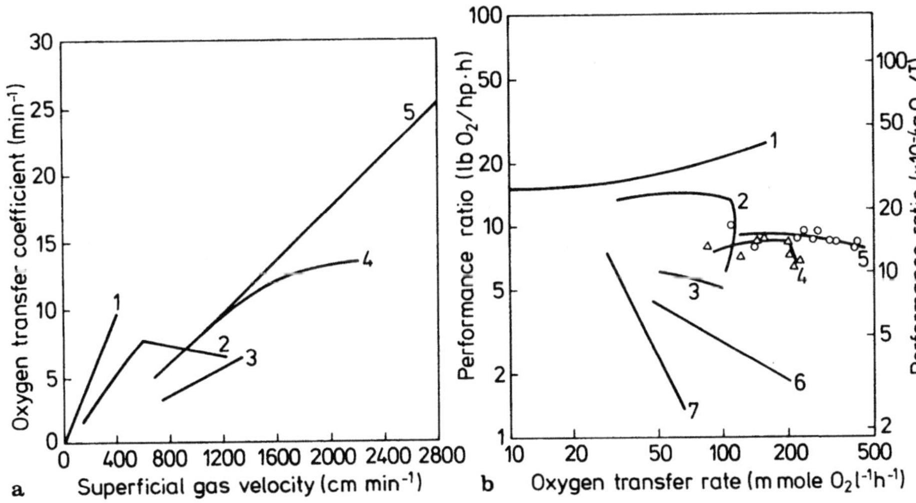

Fig. 15a. Oxygen-transfer coefficient as a function of superficial gas velocity: 1, rectangular airlift; 2, one-stage split-cylinder airlift; 3, concentric cylinder airlift; 4, one-stage split-cylinder airlift; 5, two-stage split-cylinder airlift [78]; **b**: Performance ratio as a function of oxygen-transfer rate: 1, rectangular airlift; 2, one-stage split-cylinder airlift; 3, concentric cylinder airlift; 4, one-stage split-cylinder airlift; 5, two-stage split-cylinder airlift; 6, agitated tank, 7, aeration tower [78]

group of Schügerl have done extensive modeling and experimental measurements on airlift reactors with external loop, from the bench-top scale through a 40-meter production reactor. Their approach has been essentially to measure the relevant variables, e.g. oxygen, substrate, and biomass concentration profiles, and through statistical analysis fit the results to their models. In this way they have derived correlations for key parameters that are valid under actual experimental conditions for their particular reactor system. Listed in the table are their results for the volumetric mass transfer coefficient from one of their studies [93].

Deckwer et al. [32] have presented one of the few theoretical and experimental studies on the heat transfer process. Employing the surface renewal theory for interfacial mass transfer and Kolmogorov's isotropic turbulence model, they derived and experimentally tested the results which are included in the table.

In summary, it is apparent from the system sizes indicated in the table that most experimental data have been obtained from small laboratory reactors. This means that it would be questionable to use these results in scale-up studies since it would be difficult to maintain hydrodynamic similarity between these systems and large units. Another factor is that it is difficult to compare the results of one experiment with those of another because all relevant hydrodynamic conditions are not indicated in every case. One such key variable is the ratio between the cross-sectional areas of the riser and the downflow section. Merchuk and Stein [70] in their study appropriately emphasized the importance of this parameter in the hydrodynamics of the airlift reactor.

2.3 Immobilized Biocatalyst Reactors

Immobilized bioreactors are heterogeneous solid phase-fluid phase reactors in which the biocatalyst is attached to a solid support, encapsulated in beads, or entrapped in a confined space. The biocatalyst may be immobilized as free enzymes or whole cells which contain the active enzymes. The substrate, usually carried in an aqueous medium, is passed by the anchored catalyst at rates that can be several times higher than those encountered with free biocatalysts in, for example, continuous flow stirred tank reactors (CFSTR). The potential for significant increases in throughput is one major attraction of immobilized biocatalyst reactors, since this would translate into significant reductions in reactor volume. Immobilized biocatalyst systems have been utilized with all traditional reactor configurations: BSTR, CFSTR, fixed-bed, fluidized-bed, and tubular, such as hollow-fiber biochemical reactors which will be discussed in this section.

Traditional chemical reactor engineering provides a vast repertoire of analytical techniques which may be adapted to the modeling of immobilized biocatalyst reactors [4]. However, the adaptation is not usually straightforward because the process of biocatalyst immobilization may superimpose several extraneous phenomena (relative to diffusional mass transfer) on the inherent biochemical kinetics. Other physical processes may have to be considered in addition to species mass transfer, and heat transfer problems which traditional chemical reactor engineering has essentially solved.

There are four major methods for immobilizing biocatalysts, namely:

a) Crosslinking within or on a suitable carrier;
b) Covalent bonding to a polymeric matrix;
c) Adsorption on or in a carrier, and
d) Encapsulation or entrapment.

Both enzymes in vitro and whole cells (both viable and lysed) have been successfully immobilized [72]. Method (d) applies especially to fibers and membranes with the hollow fiber reactor being a major example. The supports used for immobilization belong to two main classes of material, that is: non-porous and porous. Members of the group of porous carriers include: a) controlled pore materials which are characterized by a narrow pore-size distribution, e.g. controlled pore glass (CPG), b) broad-pore distribution carriers, e.g. alumina and c) gel structures.

Depending on the mode of immobilization and the carrier employed, physical phenomena may be superimposed on the inherent biochemical kinetics which may perturb the analysis provided by traditional catalytic chemical reaction engineering. Some of these extraneous phenomena are [103a)]:

a) Conformational or steric effects which upset the normal three-dimensional configuration of the catalysts required for its activity (in the case of enzymes). This could lead to outright denaturation or deactivation of the catalyst.
b) Partition effects which could arise from the use of electrically charged supports and/or substrates. This upsets the pH dependence of the catalyst (enzyme) and introduces an electric field component to the normal diffusional mass transfer mechanism. In order to accommodate this higher level of complexity appropriate electrostatic theory has to be incorporated in the analysis.

The relative importance of each type of extra complication in a given situation will depend on both the mode of immobilization and the type of carrier material. It can be seen therefore that immobilized biocatalyst reactors can be difficult to model. For any meaningful attempt of system conceptualization there is the need to specify a priori,

a) the immobilization mode;
b) the type of carrier; and
c) the type of biocatalyst-enzyme or cells.

Typical modeling assumptions for immobilized biocatalyst reactors include:

a) plug flow or Poiseuille-type velocity profiles;
b) steady state;
c) first-order irreversible reaction or Michaelis-Menten or substrate inhibition kinetics;
d) isothermal operation;
e) no time delay in case of recycle.

The specific forms of the governing equations and boundary conditions depend on the type of immobilization and the structure of the carrier. The literature treats:

1) Non-porous carrier and physically adsorped catalyst:
   In this case, the biochemical reaction is coupled with external film diffusion. For charged carrier and/or substrate, electrostatic effects may be incorporated.
2) Porous carrier and any of catalyst immobilization:
   a) reaction is coupled with external and possibly intra-particle mass transfer;
   b) case (a) with electrostatic effects;
   c) case (a) with catalyst deactivation;
   d) case (c) plus electrostatic effects, i.e., reaction coupled with external and intro-particle mass transfer, catalyst deactivation and electrostatic effects.

2.3.1 Hollow Fiber Biochemical Reactor (Case Study)

The key feature of hollow fiber biochemical reactors (HFBR) is that the biocatalysts (enzymes or cells) are separated from the reactant by a thin membrane (Fig. 16).

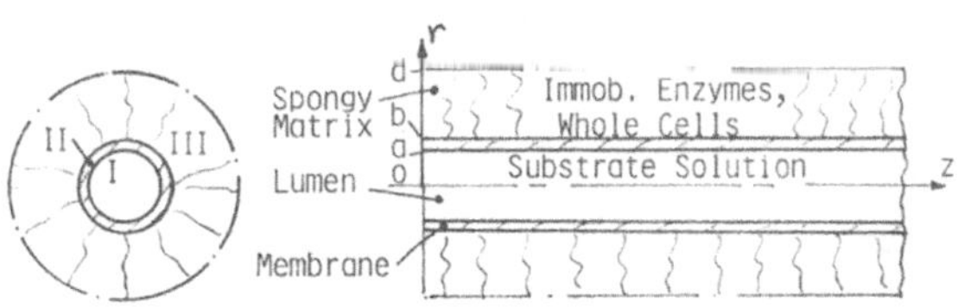

Fig 16. Schematics of single hollow fiber

The membrane permits passage of substrate and products but is impermeable to microorganisms. Entrapment of the catalysts, usually in the annular region, can be accomplished by using: a carrier liquid in a spongy matrix, just a carrier liquid possibly as a counter-current to the feed stream in the lumen, and a porous matrix for enzyme immobilization. Several geometries are utilized: planar, cylindrical (commonly used), and spherical (micro-encapsulation).

The advantages of HFBR's are: little steric hindrance and deactivation for enzymes, high concentration of microorganisms, high volumetric productivity, basically no washout, and possibility of using multi-culture conversion. Problems with this type of reactor are primarily associated with the membrane, such as fouling, leakage, and diffusion limitations for substrates and products. In addition, complications with the control of cell growth and substrate transfer resistance into the cell mass can be expected. Hence, the achievement of a continuous, high-volume concentrated product stream is still a research goal. *Experimental work* with whole cells in hollow fiber reactors has been reported by Waterland et al. [113], Kim and Cooney [55], Mattiasson and Ramstorp [69], Roy et al. [87] and Inloes et al. [53, 54]. The results by Waterland et al. [113] were used to calibrate the Kim & Cooney model for our simulation study discussed below. Roy et al. [87] used a HFBR with different fiber packing in a batch recycle mode and a continuous flow apparatus. The results in terms of lactic acid production and substrate uptake (Fig. 17) were compared with the performance of

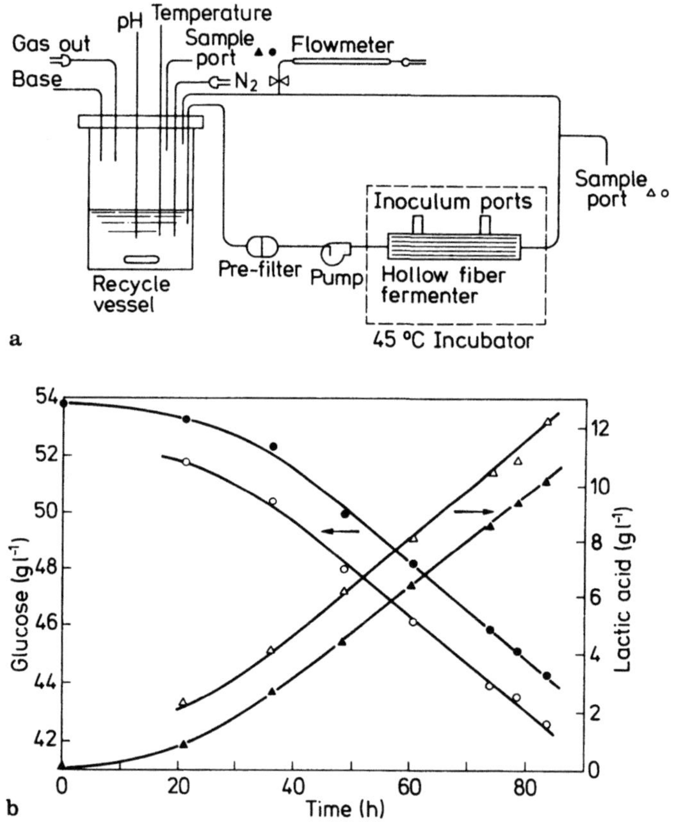

Fig. 17a. Hollow fiber batch recycle apparatus [87]; **b.** 60 Fiber HFBR module; operational mode and symbols given in Fig. 17a [87]

a CSTR operated in a batch mode (Table 4). Clearly, the high volumetric productivities of lactic acid up to 138 g l^{-1} h^{-1} obtained, is a definite performance advantage of the HFBR concept. On the other hand, specific productivity, i.e. g product dry cell weight per h and the yield per g in terms of g DW cells per substrate are in favor of the batch reactor. In addition, the problems of excessive cell growth leading to fiber (bundle) contortion and diffusion limitations have to be solved in the near future. Mathematical modeling and process simulation can aid in finding solutions

Table 4. Comparison of two hollow fiber reactors to batch growth

Number of fibers	60	108	Batch
Fiber length, cm	5.8	5.5	—
Surface area, cm^2	16.4	28.0	—
Shell side volume, cm^3	3.4	3.2	—
Max lactic acid prod. rates, $g\ l^{-1}\ h^{-1}$			
based on shell side vol.	29	50	—
based on vessel vol. (3.6 cm^3)	28	44	5[a]
Specific productivity			
$\frac{\text{g lactic acid}}{\text{g DW h}}$	.25	—	2.0[a]
Final cell density $\frac{\text{g DW}}{\text{L}}$			
based on shell side vol.	117	—	9
based on vessel vol.	111	—	—
Yields $\frac{\text{g lactic acid}}{\text{g glucose}}$	.9 ± .1	.95 ± .1	.9 ± .05
$\frac{\text{g DW cells}}{\text{g glucose}}$	.055	—	.19
flow rate, $ml\ h^{-1}$	60	65	—

[a] In exponential phase, on identical medium. This compares well to the value quoted by Friedmann and Gaden [37].

to these problems. Huffman-Reichenbach and Harper [52] tested β-galactosidase retention and activity in hollow fiber bioreactors (polysulfone membranes) under different operational conditions (Fig. 18). They reported: (1) 32% loss of enzyme activity within 15 min in the ultrafiltration mode but minimal leakage; (2) 1.2% enzyme leakage in the recycle mode for flow rates less than 200 ml per min but 30% leakage for flow rates greater than 6000 ml per min; (3) 40% leakage in the backflush mode; (4) addition of bovine serum albumin (BSA) to a buffer solution reduced the loss of free enzyme activity; and (5) 0 to 13% enzyme leakage in the static mode. It is apparent that in many cases membranes no longer can be treated as simple mechanical sieves separating solutes according to different molecular weights. In general, it is important to consider the mode of reactor operation and the type of solute, solvent, and membrane since enzyme retention characteristics depend in particular on the diffusion rate and convective flux, fluid shear stresses, axial flow rate and enzyme characteristics (concentration, shape, size, deformability, etc.). There is evidence that the size of the "apparent" membrane pores and the corresponding enzyme retentions are controlled by a dynamic relationship between interactions of membrane-flux-flow-solvent-solute. For example, certain solvents may cause membrane swelling that alters the "apparent" pore size or changes in enzyme shape and elasticity. In addition, radial volume flux may modify the membrane structure leading to solute leakage at higher flux values. Hence in cases where leakage is significant, a dynamic membrane theory has to replace the simple algebraic expressions for volume and mass flux across membranes based on irreversible thermodynamics [58].

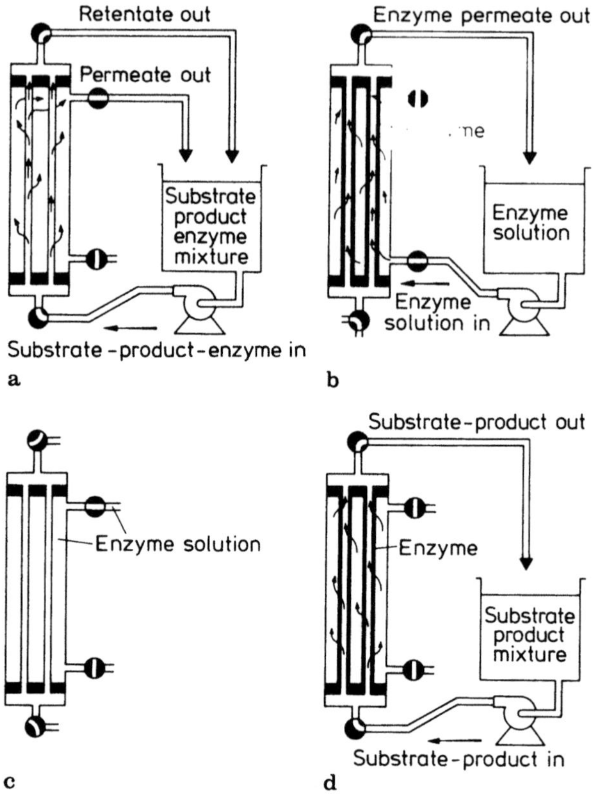

Fig. 18a–d. Schematic of hollow fiber membrane cartridges showing four modes of operation [52]; **a** Ultrafiltration mode; **b** Backflush mode; **c** Stationary mode; **d** Recycle mode

There have been a number of mathematical models published simulating substrate convection, diffusion and uptake in hollow fiber bioreactors. Rony [86] presented the first solution for immobilized enzymes. His work was advanced with contributions made by Waterland et al. [113], Kim and Cooney [55], and Webster and Shuler [114a]. Webster et al. [115] presented a solution for entrapped whole cells evenly distributed in the annular region. Recently, Webster and Shuler [114b] suggested a transient diffusion model to calculate the radial substrate concentration profiles. Kim and Cooney's model together with the transient simulator by Webster and Shuler [114b] is reviewed below. All these state-of-the-art models have certain features in common. The modeling equation used is a reduced linear form of the mass transport Eq. (10) given in Sect. 2.1. Product kinetics are not considered although under the assumption of a Gaden type-I reaction, the product concentration profiles could perhaps be deduced from substrate curves. The difference between the kinetics of free or pseudo-immobilized enzymes and entrapped whole cells is not addressed. Even if it is assumed that cell growth is terminated before the microorganism are entrapped, the simplified Michaelis-Menten expression used is hardly applicable. The membrane functions as an ideal semi-permeable barrier where concentration polarization and (radial) volume fluxes have been neglected. The simplified modeling equtions are solved analytically.

A representative hollow fiber of which hundreds may be packed into one tube, and hollow fiber bioreactors are schematically shown in Figs. 16, 17a and 18. A laminar feedstream in the lumen (Region I) is a continuous source of substrate molecules which diffuse through the thin membrane (Region II) into the stagnant annular region or spongy matrix (Region III) in which the microbial catalysts are well distributed. The governing equation for substrate transport and uptake in hollow fiber bioreactors can be directly deduced from Eq. (7) as:

$$\frac{\partial c}{\partial t} + v_z \frac{\partial c}{\partial z} = \frac{D_r}{r} \frac{\partial}{\partial r}\left(r \frac{\partial c}{\partial r}\right) - kc\,. \tag{41}$$

Kim and Cooney [55)] considered the steady state version of (41) whereas Webster and Shuler [114b)] solved Eq. (41) without the convection term. The sink term, kc, i.e. biochemical conversion of the substrate, is a linearized expression of the Michaelis-Menten function where $k = V_{max}/K_m$. Obviously, radial convection of substrate, $v_r\,\partial c/\partial r$ and axial diffusion $D_z\,\partial^2 c/\partial z^2$ have been neglected based on a relative order of magnitude analysis.

With the system conceptualization depicted in Fig. 15 and the assumptions discussed earlier, Kim and Cooney [55)] postulated the following modeling equations and associated boundary conditions

$$\text{(lumen)} \quad \frac{D_1}{r} \frac{\partial}{\partial r}\left(r \frac{\partial c_1}{\partial r}\right) = v_0 \left(1 - \frac{r^2}{a^2}\right) \frac{\partial c_1}{\partial z} \tag{42}$$

$$0 < r < a\,; \qquad 0 < z < L$$

$$\text{(membrane)} \quad \frac{D_2}{r} \frac{\partial}{\partial r}\left(r \frac{\partial c_2}{\partial r}\right) = 0, \qquad a < r < b \tag{43}$$

$$\text{(spongy matrix)} \quad \frac{D_3}{r} \frac{\partial}{\partial r}\left(r \frac{\partial c_3}{\partial r}\right) = \frac{V_{max}}{K_m} c_3\,; \qquad b < r < d \tag{44}$$

The boundary conditions are:

$$\left.\begin{aligned}
&\text{at}\ \ \underline{r = a}: \ -D_1 \frac{\partial c_1}{\partial r} = -D_2 \frac{\partial c_2}{\partial r}\,; \qquad \frac{c_2}{c_1} = K_a \\
&\text{at}\ \ \underline{r = b}: \ -D_2 \frac{\partial c_2}{\partial r} = -D_3 \frac{\partial c_3}{\partial r}\,; \qquad \frac{c_2}{c_3} = K_b \\
&\text{at}\ \ \underline{r = d}: \ D_3 \frac{\partial c_3}{\partial r} = 0\,.
\end{aligned}\right\} \tag{45}$$

Clearly, the boundary conditions (45) which couple the set of modeling equations, impose a particular behavior on the system which does not necessarily coincide with the actual process dynamics. On the other hand, the radius d could be selected such that there is no substrate flux at r = d, i.e. the solution is contained within the hollow fiber subunit.

Equation (42) is solved using separation of variables and infinite series method to yield:

$$C_1(X, Z) = \sum_{n=2}^{\infty} A_{1n} \exp(-\Lambda_{1n}^2 Z) \exp(-\Lambda_{1n} X^2/2)\, M\left(\frac{2 - \Lambda_{1n}}{4}, 1, \Lambda_{1n} X^2\right). \tag{46}$$

Equation (43) can be directly integrated yielding a logarithmic function and Eq. (44) is actually an O.D.E. for $c_3(r)$ whose solution is:

$$C_3(X, Z) = \sum_{n=2} A_{3n} \exp(-\Lambda_{1n}^2 Z)\, [I_0 X] + \frac{I_1(\lambda\beta)}{K_1(\lambda\beta)} K_0(\lambda X). \tag{47}$$

The *dimensionless variables* are defined as:

$$C = \frac{c}{c_0}, \quad X = r/a \quad \text{if} \quad r < a \quad \text{and} \quad X = r/b \quad \text{if} \quad r > b;$$

$$Z = \frac{z}{aPe}, \text{ where } Pe = \frac{v_0 a}{D_1} \text{ is the Peclet number.}$$

The eigenvalues Λ_{1n} are the roots of

$$\left[\sigma \frac{dF_1}{dX} + F_1\right]_{X=1} = 0 \quad \text{for} \quad n = 2, \infty$$

where $F_1 = \exp[-1/2X^2]\, M\left(\frac{2 - \Lambda}{4}, 1, X^2\Lambda\right)$ and M is the Kummer function.

The *system parameters* are defined as:

a) $\sigma = \frac{1}{h} - \frac{K}{\alpha} \left.\frac{F_3}{dF_3/dX}\right|_{X=1}$ an overall mass transfer resistance;

where $h = -K_a \frac{D_2}{D_1} \frac{1}{\ln a/b}$ is a (very large) membrane parameter;

$K = \frac{K_a}{K_b} = \frac{c_3}{c_1}$, $\alpha = \frac{D_3 b}{D_1 a}$ also depend on the membrane characteristics;

and $F_3 = I_0(\lambda X) + \frac{I_1(\lambda X)}{K_1(\lambda X)} K_0(\lambda X)$.

b) $\lambda^2 = \frac{b^2 V_{max}}{D_3 K_m}$ is the Thiele modulus squared (ratio of bioreaction to diffusion phenomena or τ_{diff}/τ_{react}).

Note: $\lambda^2 < 1 \rightarrow$ kinetic control; $\lambda^2 > 10^3 \rightarrow$ diffusion control; at intermediate values. enzyme activities will be reduced by diffusion effects.

c) $\beta = d/b$ is a geometric factor reflecting the ratio of enzyme to substrate volume. The *coefficients of the infinite series* are defined as:

$$A_{1n} = -\left[\frac{2}{\Lambda} \frac{\frac{dF_1}{dX}}{-\frac{d}{d\Lambda}\left(\frac{dF_1}{dX}\right) + \frac{dF_1}{d\Lambda}\frac{dF_1}{dX}}\right]_{X=1,\Lambda=\Lambda_{1n}} \quad \text{where } n = 2, \infty$$

and

$$A_{3n} = \left[\frac{dF_1}{\alpha(dF_3/dX)} \frac{dF_1}{dX}\right]_{X=1,\Lambda=\Lambda_{1n}} \quad \text{where } n = 2, \infty$$

The solution procedure starts with the selection of appropriate values for λ, β, K and α in order to evaluate σ which for h $\gg$ 1 or $h^{-1} \approx 0$ reads:

$$\sigma = \frac{K}{\alpha} \frac{I_0(\lambda) K_1(\lambda\beta) + I_1(\lambda\beta) K_1(\lambda)}{\lambda[I_1(\lambda\beta) K_1(\beta) - I_1(\lambda) K_1(\lambda\beta)]} \tag{48}$$

σ and α are needed to determine Λ_{1n}, A_{1n}, and A_{3n}. The eigenvalues Λ are found by a search over intervals as indicated by Kim and Cooney [55], to yield crude guesses which are then refined with a special I.M.S.L. Fortran library routine. The Kummer function is evaluated noting that:

$$M(a, b, y) = 1 + \frac{ay}{b} + \sum_{n=2}^{\infty} T_n \quad \text{where} \quad T_n = \frac{a + n - 1}{b + n - 1} T_{n-1} .$$

The derivative $\frac{dF_1}{dX}$ can be written in terms of exponentials and Kummer functions, thus they are computed "exactly". However the analytical expressions for $\frac{dF_1}{dX}$ and $\frac{d}{d\Lambda}\frac{dF_1}{dX}$ are cumbersome; therefore following Kim and Cooney's procedure they are evaluated numerically using a first order back difference formula where $\Delta\Lambda = 10^{-6}$. It was only necessary to locate the first five eigenvalues to compute solutions to excellent accuracy. Double precision was used everywhere. Thus, Kim and Cooney's claim that their solution is better than the solution of the same model proposed by Waterland et al. [113] is correct to some degree. However, Waterland's method allows for an axial variation of the rate constant (which actually occurs with some enzymes) by employing a slicing technique.

The model is calibrated by comparing computer simulation results with experimental observations, including error margins, reported by Waterland et al. [113]. Figure 19 shows the change of substrate bulk concentration vs. Thiele modulus for various stations along the axial direction or different flow rates. The dimensionless membrane group $D_1/\lambda D_2 = 10$. Model input data taken from Waterland et al. [113] are listed in the Appendix. The bulk or mixing cup concentration is defined as:

$$\bar{C}_1(z) = \sum_{n=2}^{\infty} B_{1n} \exp(-\Lambda_{1n}^2 z) \tag{49}$$

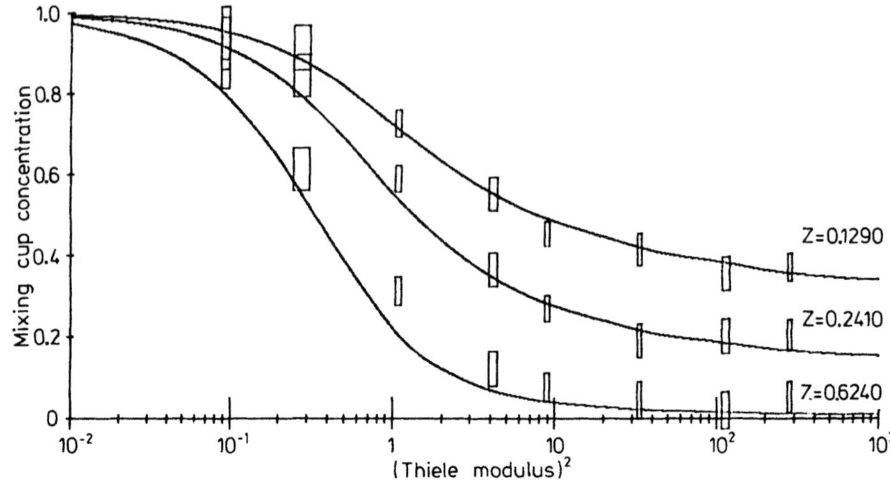

Fig. 19. Mixing cup concentration vs. Thiele modulus squared with parameter Z ~ z or Q. (Measured data w/error bars from Waterland et al. [113])

$$\text{where} \quad B_{1n} = \left[-\frac{4A_{1n}}{\Lambda_{1n}} \frac{dF_1}{dX} \right]_{X=1} .$$

A slight discrepancy between model prediction and data points is apparent at $\lambda^2 = 1$, especially downstream, e.g. $Z = 0.6240$. It is interesting to note that a small increase in σ, a parameter which reflects membrane characteristics, would result in a better fit.

For a comprehensive sensitivity analysis, the following cases were investigated:

a) Effectiveness factor, η, vs. Thiele modulus squared, λ^2, for three different β-values, keeping $K/\alpha = 1.0$ constant (Fig. 20). The effectiveness factor η can be defined

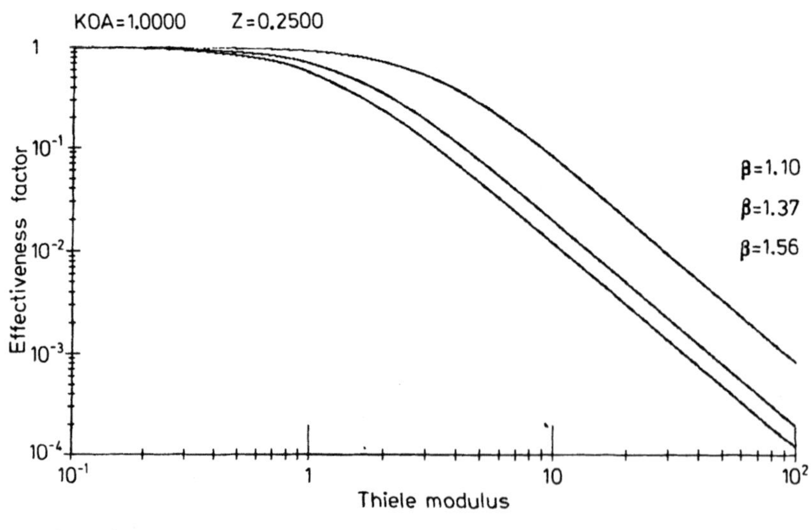

Fig. 20. Effectiveness factor vs. Thiele modulus for given K/A and Z (parameter = β)

as the ratio between the amount of substrate that actually reacted and the amount that could have reacted if all biocatalysts were exposed to substrate concentration at inlet. Mathematically,

$$\eta = \frac{1 - \bar{C}_1(z)}{(C_0 - C_f)/C_0} = [1 - \bar{C}_1(z)]/[2\lambda^2 z(\beta^2 - 1)\, D_3/D_1] \tag{50}$$

where the bulk substrate concentration $\bar{C}_1$ is given above.

The β-values used (i.e. β = 1.10, 1.37 and 1.56) reflect commercially available fibers. A number of conclusions can be drawn from Fig. 20. The effectiveness factor is almost independent of λ, z, and β when the Thiele modulus is quite small, i.e. the process dynamics are kinetically controlled. With increasing λ, diffusional effects become dominant and η decreases. The same functional behavior has been observed for biocatalyst beads [57]. Specifically, η decreases with growing β, z and λ. Diffusional mass transfer in the lumen becomes so limited that less substrate is available for bioconversion by the microorganisms in the spongy matrix.

b) Effectiveness factor vs. Thiele modulus at constant β and with variable K/α, i.e. K/α = 0.9, 1.0 and 1.1. Since no significant effect for K/α ≠ 1.0 was detected, only one plot is shown for Z = 0.1300 in Fig. 21.

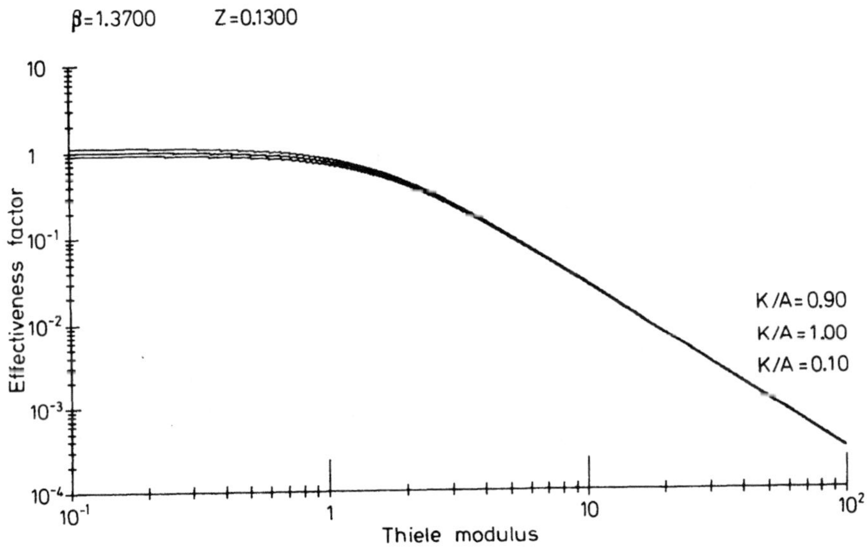

Fig. 21. Effectiveness factor vs. Thiele modulus for given β and Z (parameter: K/A)

c) Radial substrate concentration profiles for two different λ-values (λ_1 = 1.0 and λ_2 = 10.0) and constant β = 1.75 and K/α = 1.0 (Figs. 22a, b). Although b/a = 1.005 indicates a very thin membrane, the concentration drop across the membrane is detectable.

Conclusions:

- At higher λ (tending toward diffusional control) substrate molecules do not penetrate the matrix very far and even less as Z increases. Also the concentration

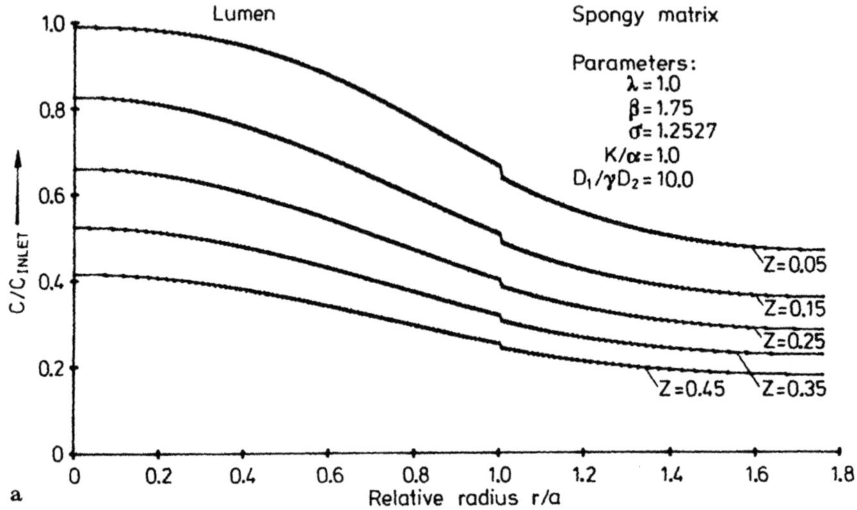

Fig. 22a. C(R) for some Z values: hollow fiber model

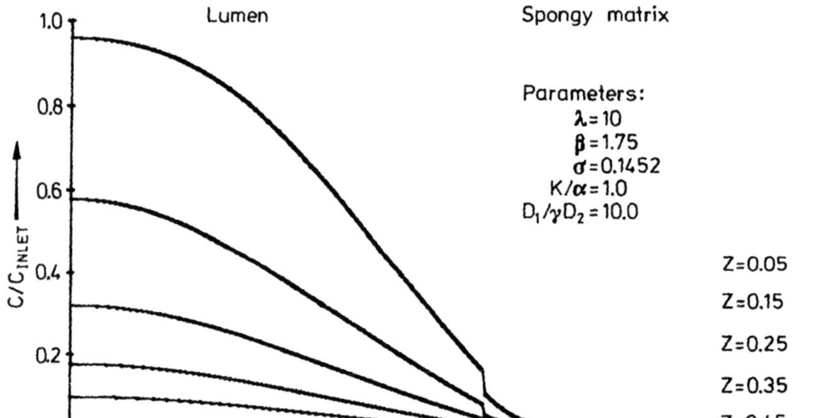

Fig. 22b. C(R) for some Z values: hollow fiber model

at the membrane is very low, however, the gradient is higher than for low λ. The resistance to mass transfer seems to be in the lumen, predominantly. Thus, models which assume perfect mixing in the lumen are grossly inaccurate. Enzymes far from the wall are not contacted by substrate.

- At higher λ the variation of C with Z is stronger. This is reasonable as substrate conversion in the reactor occurs at a higher rate.
- At lower λ the resistance to mass transfer seems roughly equally distributed between lumen and matrix. Thus, with slower-growing microorganisms the substrate

molecules spend more time traveling in the matrix before being consumed, than with a higher λ. Once again significant radial gradients exist.
- At lower λ substrate manages to exist at the outer edge of the matrix.

d) Overall mass transfer resistance, σ, vs. Thiele modulus, λ (Fig. 23). The conditions are $K/\alpha = 1.0\ l\ h^{-1} = 0$ and $\ln a/b = 0$ while the parameter β is varied as $\beta = 1.10$, 1.37, and 1.56.

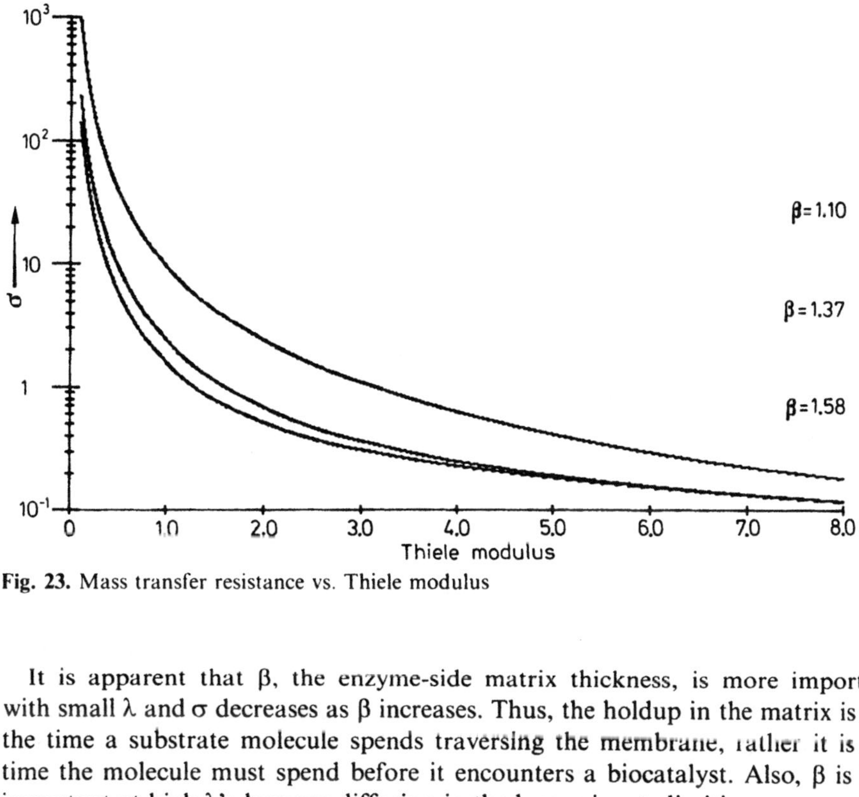

Fig. 23. Mass transfer resistance vs. Thiele modulus

It is apparent that β, the enzyme-side matrix thickness, is more important with small λ and σ decreases as β increases. Thus, the holdup in the matrix is not the time a substrate molecule spends traversing the membrane, rather it is the time the molecule must spend before it encounters a biocatalyst. Also, β is less important at high λ's because diffusion in the lumen is rate limiting.

Webster and Shuler [114b)] developed a model simulating transient radial diffusion of substrate with irreversible first-order reaction kinetics. Of the three regions discussed earlier (Fig. 16) only Region III, the spongy matrix is considered. Hence, the tailored equation can be directly obtained from Eq. (41) as

$$\frac{\partial c_3}{\partial t} = \frac{D}{r}\frac{\partial}{\partial r}\left(r\,\frac{\partial c_3}{\partial r}\right) - kc_3; \qquad b < r < d \tag{51}$$

with initial condition

$$C_3(r, 0) = 0 \tag{52}$$

and boundary conditions

$$c_3(r = b) = c_0 \quad \text{and} \quad \left.\frac{\partial c_3}{\partial r}\right|_{r=d} = 0 \tag{53}$$

Note, that $b = R$ and $d = \beta R$ in the original article. The typographical errors in the paper, especially the placement of parentheses in the solution $c_3(r, t)$ made it necessary to rederive the solution. Typical results are given in Fig. 24 showing substrate depletion in space and time within the annular region.

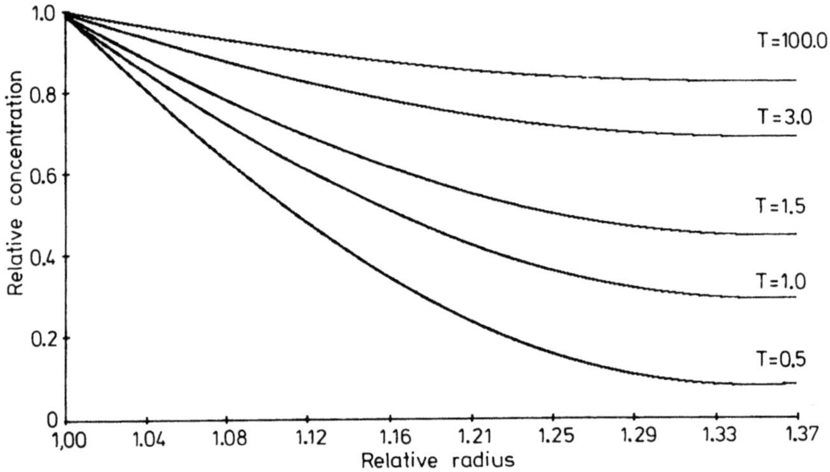

Fig. 24. Dynamic simulation of a hollow fiber reactor [114b)]

Although axial concentration gradients are ignored, Webster and Shuler [114b)] indicated that this is justified since the hollow fiber unit was installed on a recirculating loop in a CSTR. In this mode of operation, in contrast to using the module as a small "plug flow" reactor, the variations in the z-direction may be less than in the r-direction and the axial dynamic effects in the lumen may be much faster than in the r-direction within the spongy matrix. Furthermore, high fluid velocities in the lumen would make the radial diffusion effects across the membrane and in the spongy matrix more important than radial diffusion in the lumen. However, this model ignores the membrane; since the membrane is selectively permeable it may provide significant resistance to diffusion. Thus, the transient effect of the membrane might be significant. The transient model is computationally convenient, however profiles at time nearly zero exhibited an erratic behavior. When the series solution was evaluated to more than the 20 terms, these irregularities vanished. The eigenvalues depend on β and R but not on D, V_{max} or K_m; thus sensitivity analyses can be easily performed. Furthermore the solution can be integrated to yield substrate consumption vs. time profiles. These are much more amenable to experimental verification.

In general, it can be concluded that hollow fiber biochemical reactors have great potential to convert substrate (e.g. glucose solutions) to highly valuable products. The basic reactor designs can be enhanced to obtain improved flow dynamics (e.g. radial flow across the membrane) and product separation which should result in better reactor performance. Consequently, future modeling studies will address: the complete mass transfer equation with improved dynamics, non-isothermal operations, more realistic biokinetics, including cell growth, and the details of trans-membrane fluxes. However, with regard to the last point, Waterland et al. [113] stated that the membrane region (thickness 0.5 μm) presents negligible diffusional resistance to the reactor operation. As demonstrated earlier (Fig. 22) this is not the case if λ is sufficiently large.

2.3.2 Fixed-bed and Fluidized-bed Bioreactors

Another class of immobilized biocatalyst reactors comprise the fixed-bed and fluidized-bed bioreactors. Their unique feature is that the biocatalyst is immobilized on a support which is large compared to microbial (or enzyme) dimensions, e.g. sand, controlled pore-glass, and glass, ceramic or polymer beads. These reactors are similar to those employed in traditional chemical process engineering and the methods for their analysis derive substantially from current chemical reaction engineering techniques. However, major points of difference include: the choice of the fluid phase, catalyst deactivation mechanisms, and the possibility of the superposition of electrostatic effects on the inherent kinetic and mass transfer processes. The fluid phase is predominantly liquid, especially in two-phase fixed-bed and fluidized-bed bioreactors. It is in principle possible to use three phases, i.e. gas-liquid-solid bioreactors, as reported, for example, by Fan et al. [37a, b]. In such a system, the liquid phase would carry the substrate and products, and the gas (usually air) phase would provide the oxygen needed for the reaction (aerobic processes only). Catalytic deactivation is due mainly to hydrodynamic shear effects; the biocatalyst, in the form of a film is either washed off the support, or the shear causes enzyme inactivation [23, 103]. The third main difference is the presence of electrostatic effects in the mass transfer processes, especially when porous supports are used for biocatalyst immobilization. These arise from charge distributions on the substrate and/or the surface of the support.

Typical geometrical arrangements for fixed-bed and fluidized-bed bioreactors are indicated in Fig. 25. Reactor geometry is predominantly the cylindrical tube with a constant cross-sectional area. This arrangement may be used in either the fixed-bed or fluidized-bed mode, and with the possibility of recycle as indicated. Recent studies have however, investigated the use of tapered cylindrical tubes for the fluidized-bed application [51, 94]. Tapering the column accommodates a wider range of fluid flow rates for stable operation and also reduces the loss of the bed material.

In the modeling of this class of bioreactors, the key processes to be considered include mass transfer processes and the fluid dynamics, assuming that the inherent microbial/enzymatic kinetics are well understood. The analysis of mass transfer depends on the structure of the biocatalyst carrier (porous or nonporous) and the mode of biocatalyst immobilization. With nonporous carriers, only external film diffusion is considered. Electrostatic effects may in addition be considered if the carrier and/or the substrate is not electroneutral. In either case, the method of analysis is similar to that for external transport in heterogenous reaction system [98]. When a porous carrier is used, for example, controlled-pore glass or any of the several biopolymers, both intra-particle and external mass transfer effects have to be con-

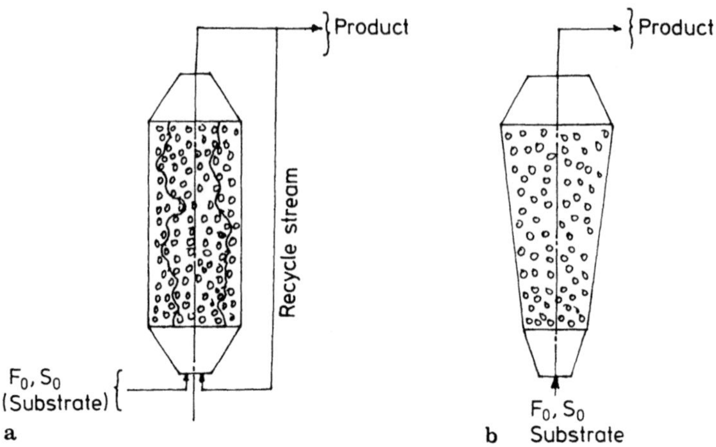

Fig. 25a. Tubular fixed-bed (or fluidized-bed) bioreactor; **b** Tapered fluidized-bed bioreactor

sidered [57]. The method of analysis is similar to that for mass transfer with reaction in synthetic porous catalysts [98] modified if necessary by the presence of electrostatic effects.

The flow field for the fluid is assumed to be ideal plug flow, or dispersed plug flow. In the latter case an empirical axial dispersion coefficient is introduced to describe the departure of the flow from ideality (plug flow). For steady flow, and neglecting all radial variations, the governing equation for the substrate for example, would be:

$$D_{ax}\frac{d^2[S]}{dz^2} - \bar{u}\frac{d[S]}{dz} - \frac{r_s}{\varepsilon} = 0,$$

where D_{ax} is the empirical axial dispersion coefficient, [S] is the substrate concentration, r_s is the substrate consumption rate, and z is the axial distance. $\bar{u}$ is the superficial fluid velocity and ε is the bed porosity.

Both the fixed-bed and fluidized-bed bioreactor have been analyzed using equations like that above in several published works. For both types of reactor, the following aspects are discussed in the literature:

- biocatalyst immobilization [2];
- biocatalyst deactivation [34, 110];
- waste treatment [101];
- starch hydrolysis [2, 35];
- electrostatic effects [43, 96a, 96b];
- mass transfer effects [37, 63, 84, 85, 106];
- dispersion for packed-bed reactors [21];
- hydrodynamic characteristics [7, 37];
- steady state analysis of fluidized-bed bioreactors [3].

A computer simulation study of fluidized-bed bioreactors is not warranted due to a lack of comprehensive and reliable data sets.

3 Conclusions and Future Work

Thoenes [102] concluded in a review of current modeling problems of chemical reactors, that in many cases the limits of pure physical modeling have been attained. There appears to be an increasing need for mathematical modeling studies concerning the simultaneous effects of several, often transient and nonlinear, phenomena that occur in real reactors. However, existing computer models fall short in simulating the reactor process dynamics with sufficient resolution, accuracy, and completeness. Hence, the following problem areas should be addressed, employing computer simulation models and experimental data sets for model calibration/verification:

- transient, non-isothermal process dynamics;
- interfacial phenomena in gas-liquid and gas-liquid solid flows;
- biochemical reactions, catalysis and cell growth;
- rheology of non-Newtonian fluids;
- turbulent two- and three-phase flow;
- mass transfer in complex flows;
- product recovery and by-product separation;
- biochemical process control;
- parameter estimation;
- innovative design of practical bioreactors; and
- scale-up from bench-top reactors and optimization of large-scale bioconversion systems.

4 Acknowledgement

The authors wish to thank Professor H. Bungay for his support, Jeff Schonberg for his computational work, and the speakers as well as participants of RPI's annual "Advanced Biochemical Engineering" seminar for their input. Special thanks go to Ms. Lorraine McGrath who did an excellent job in wordprocessing the manuscript.

5 Appendices

5.1 Appendix AF: Nomenclature and Input Data for Airlift

Nomenclature (Sect. 2.2.1)

A_d	cross-section of the downcomer (cm^2)
A_r	cross-section of the riser (cm^2)
B	biomass concentration ($mg\ l^{-1}$)
B_A	dimensionless biomass concentration, $B/Y\ S_f$
c	oxygen concentration in the liquid phase ($M\ mol\ l^{-1}$)
c^*	oxygen concentration in the liquid in equilibrium with air ($M\ mol\ l^{-1}$)

D	dilution rate (s^{-1})
D_1	stoichiometric group, $= \frac{Y_s S_f}{Y_{ox} M_{ox}}$ (dimensionless)
D_G	oxygen diffusivity in the gas phase ($cm^2\ s^{-1}$)
D_L	oxygen diffusivity in the liquid phase ($cm^2\ s^{-1}$)
F	volumetric flow rate ($cm^3\ s^{-1}$)
H	Henry's constant (liter atm/mol)
J	average volumetric flux density of the mixture ($cm\ s^{-1}$)
J_G	gas volumetric flux density ($cm\ s^{-1}$)
J_L	liquid volumetric flux density ($cm\ s^{-1}$)
K	dimensionless Michaelis constant, $= K_m/S_f$
K_m	Michaelis constant ($mg\ l^{-1}$)
K_La	volumetric mass transfer coefficient (s^{-1})
L	height of the riser (cm)
M_s	molecular weight of the substrate ($g\ mol^{-1}$)
M_{ox}	molecular weight of the oxygen ($g\ mol^{-1}$)
P_{tot}	total pressure (atm)
P_{O_2}	oxygen partial pressure (atm)
Q_L	liquid volumetric flow rate ($l\ s^{-1}$)
Q_G	gas volumetric flow rate ($l\ s^{-1}$)
R	reaction group, $= L\mu/J_L$ (dimensionless)
r_B	rate of biomass generation ($mg\ l^{-1}\ s^{-1}$)
r_S	rate of substrate consumption ($mg\ l^{-1}\ s^{-1}$)
r_{O_2}	rate of oxygen consumption ($M\ mol\ l^{-1}\ s^{-1}$)
S	substrate concentration ($mg\ l^{-1}$)
S_A	dimensionless substrate concentration $= S/S_f$
St_G	Stanton number for the gas phase $= (LK_La/J_G)\ (RT/H)$ (dimensionless)
St_L	Stanton number for the liquid phase $= LK_La/J_L$ (dimensionless)
t	time (s)
T	temperature (K)
V	total volume of fermentor (l)
V_G	superficial gas velocity ($cm\ s^{-1}$)
V_L	superficial liquid velocity ($cm\ s^{-1}$)
V_{SG}	gas-phase volume of the separator (l)
V_{SL}	liquid-phase volume of the separator (l)
Y_S	stoichiometric coefficient, (g) biomass per (g) substrate
Y_{ox}	stoichiometric coefficient, (g) biomass per (g) oxygen
Y_1	dimensionless partial pressure of oxygen, $P_{O_2}/P_{O_2 o}$
Y_2	dimensionless oxygen concentration, C/C^*
z	axial height (cm)
β	ratio of separator volume to total volume (dimensionless)
γ	ratio of liquid flow rates (dimensionless)
φ	gas hold-up (dimensionless) (fraction of gas volume/total volume)
η	dimensionless tube length (z/L)
μ	specific metabolic rate (s^{-1})
θ	dimensionless dilution rate $= DL/J_L$
ξ	ratio of cross-section of riser to downcomer (dimensionless)

Subscripts

d	related to the bottom of downcomer
f	related to the feed
G	gas phase
s	related to the gas separator
T	related to the top of the riser
O	related to the point of gas injection
L	liquid phase

Input Data (Sect. 2.2.1)

$T = 310$ (K)
$\mu = 0.4$ (h)$^{-1}$
$K_m = 37.5$ (mg l^{-1})
$H = 900$ (atm l^{-1} mol^{-1})
$Y_s = 0.4$ (g) biomass per (g) substrate
$L = 400$–1000 (cm)
$\beta = 0.1$
$J_G = 35$ (cm s^{-1})
$S_f = 200$–775 (mg l^{-1})
$\xi = 4.0$
$Y_{ox} = 1.0/(1.7/Y_s - 2.05)$ (g) biomass per (g) oxygen

5.2 Appendix HFBR: Nomenclature and Input Data for Hollow Fiber

Nomenclature for Kim & Cooney Model

a	inner radius of membrane
A, B	series solution constants (eigenconstants)
b	outer radius of membrane
c	substrate concentration
C	dimensionless substrate concentration
$\bar{C}$	dimensionless bulk substrate concentration
C_f	idealized effluent bulk concentration (no diffusional resistance)
D	diffusion coefficient
h	membrane mass transfer coefficient
I_0	modified Bessel function of the first kind, order zero
I_1	modified Bessel function of the first kind, order one
K_0	modified Bessel function of the second kind, order zero
K_1	modified Bessel function of the second kind, order one
K_a, K_b	membrane partition coefficients
K	membrane partition coefficient ratio
k	reaction rate constant
M	the Kummer function
r	radial position
X	dimensionless radial position

z axial position
Z dimensionless axial position

Greek Symbols

α diffusion coefficient ratio
β hollow fiber radius ratio
η effectiveness factor
λ Thiele modulus
Λ eigenvalues
σ overall mass transfer resistance

Nomenclature for Webster & Shuler Model

C_0 inlet concentration
C_3/C_0 relative concentration
D_3 effective diffusivity in spongy matrix
k first order rate constant
K_m Michaelis constant
r/b relative radius
b inner radius of spongy matrix
t time
V maximum reaction rate
$\beta = \frac{d}{b}$ relative radius of which there is no substrate concentration gradient

Input Data

Kim and Cooney (1976)

Case Study (a) (C/C_0 vs. r all regions)
$\lambda = 1.0, 10.0$
$\beta = 1.75$
$D_1/D_2 = 10.0$
$K_a = K_b = 1.0$
$\alpha = 1$
$b/a = 1.005$
$Z = 0.05, 0.15, 0.25, 0.35, 0.45$

Case Study (b) (η vs. λ, Parameter: K/α)
$\lambda = 0.1$ to 100.0
$\beta = 1.37$
$Z = 0.13$
$K/\alpha = 0.9, 1.0, 1.1$

Case Study (c) (η vs. λ, Parameter: β)
$\lambda = 0.1$ to 100.0
$\beta = 1.1, 1.37, 1.56$
$Z = 0.13, 0.25, 0.63$
$K/\alpha = 1.0$

Case Study (d) ($\bar{C}$ vs. λ^2, Parameter: Z)

$\lambda^2 = 0.01$ to 10^3
$\beta = 1.75$
$D_1/D_2 = 10.0, 100.0$
$\alpha = 1.0$
$K_a = K_b = 1.0$
$b/a = 1.005$
$z = 0.129, 0.241, 0.624$

Case Study (v) (σ vs. λ, Parameter: β)

$\lambda = 0.$ to 8.0
$\beta = 1.1, 1.37, 1.56$
$1/h = 0$

Webster and Shuler (1981)

$b = 130\ \mu m = —.013$ cm
$d/b = \beta - 1.37$
$k = V/K_m = 0.1\ s^{-1}$
$D = 6.2 \times 10^{-6}\ cm^2\ s^{-1}$
$t = 0.5$ s, 1 s, 1.5 s, 3 s, 100 s

6 References

1. Aiba, S., Humphrey, A. E., Millis, N. F.: Biochemical Engineering, 2nd edit. Tokyo: Univ. of Tokyo Press 1973
2. Allen, B. R. et al.: Biotech. Bioeng. *21*, 689 (1979)
3. Andrews, G. F.: ibid. *24*, 2013 (1982)
4. Bailey, J. E.: Chem. Eng. Sci. *35*, 1854 (1980)
5. Bailey, J. E.: Personal Communication 1981
6. Bailey, J. E., Ollis, D. F.: Biochemical Engineering Fundamentals. New York: McGraw-Hill 1977
7. Begovich, J. H., Watson, J. S.: In: Fluidization, Cambridge University Press 1978
8. Belfort, G.: In: Course Notes, Advanced Biochemical Engineering, Troy: Rensselaer Polytechnic Institute (Summer 1982)
9. Blanch, H. W., Bhavaraju, S. M.: Biotech. Bioeng. *18*, 745 (1976)
10. Blenke, H.: Adv. Biochem. Eng. *13* (1979)
11. Bowski, L. et al.: Presented at 182nd ACS Meeting, New York City, Aug. 23–28, 1981
12. Box, G. E., Jenkins, G. M.: Time Series Analysis; Forecasting and Control, Holden-Day 1976
13. Brauer, H.: Adv. Biochem. Eng. *13* (1979)
14. Bravard, J. P. et al.: Biotech. Bioeng. *21*, 1239 (1979)
15. Brodkey, R. S.: Chem. Eng. Commun. *8*, 1 (1981)
16. Bryant, J.: Adv. Biochem. Eng. *5* (1977)
17. Buhr, H. O., Andrews, J. F.: Water Research *11*, 129 (1977)
18. Bungay, H. R.: In: Enzyme engineering (Kendall Pye, Weetal, eds.), Vol. 3, Plenum Publishing Company 1978
19. Bungay, H. R.: Jour. Amer. Oil Chem. Soc. *59*, 270A (1982)
20. Calderbank, P. H.: Trans. Inst. Chem. Eng. *37*, 173 (1959)
21. Chang, H.-C.: AIChE J. *28*, 208 (1982)
22. Chang, M. M. et al.: Adv. Biochem. Eng. *20*, 15 (1982)
23. Charm, S. E., Wong, B. L.: Biotech. Bioeng. *20*, 451 (1978)
24. Christensen, D. R., McCarty, P. L.: J. W. P.C.F. *47*, 2652 (1975)
25. Cichy, P. T. et al.: Ind. Eng. Chem. *61*, 6 (1969)

26. Cohen, A. et al.: Water Research *13*, 571 (1979)
27. Collins, A. S., Gilliland, B. E.: J. Environ. Eng. Div., Proc. ASCE, 100, No. EE2, 487 (1974)
28. Cooney, C. L.: Science *219*, 728 (1983)
29. Danckwerts, P. V.: Gas-Liquid Reactions. McGraw-Hill 1970
30. Danielsson, B. et al.: Biotech. Bioeng. *21*, 1749 (1979)
31. Deckwer, W.-D.: Intl. Chem. Eng. *19*, 21 (1979)
32. Deckwer, W.-D.: Chem. Eng. Sci. *35*, 1341 (1980)
33. Deckwer, W.-D. et al.: Biotech. Bioeng. *24*, 461 (1982)
34. Do, D. D., Weiland, R. H.: ibid. *23*, 691 (1981)
35. Emery, A. N., Cardoso, J. P.: ibid. *20*, 1903 (1978)
36. Erickson, L. E., Deshpande: ibid. *24* (1982)
37a. Fan, L.-S., Hwang, S.-J., Matsuura, A.: Proc. AIChE Annual Mtg. Washington D.C., Nov. 1983
37b. Fan, L.-S., Long, T.-R., Fujie, K.: Proc. AIChE Annual Mtg. San Francisco, CA, Nov. 1984
37c. Friedmann, M. R., Gaden, E. L.: ibid. *12*, 961 (1970)
38. Goldberg, I. et al.: ibid. *18*, 1657 (1976)
39. Graef, S. P., Andrews, J. F.: AIChE Symp. Ser., *70*, No. 136, 101 (1973)
40. Gregor, H. P.: Proc. 3rd Annual Biomass Energy Syst. Conf., Golden, Colorado. June 5–7, 1979
41. Gregor, H. P., Jeffries, T. W.: In: Biochemical Engineering (Vieth et al., eds.), Vol. 326, New York: New York Academy of Sciences 1979
42. Guilbault, G.: In: Biochemical Engineering (Vieth et al., eds.), Vol. 326, New York: New York Academy of Sciences 1979
43. Hamilton, B. K. et al.: J. theor. Biol. *41*, 547 (1973)
44. Hampel, W.: Adv. Biochem. Eng. *13* (1979)
45. Hatch, R. T.: Ph. D. Thesis, M.I.T., Cambridge 1973
46. Hatch, R. T., Wang, D. I. C.: Presented at the 1st Chemical Congress of the North American Continent, Mexico City, December 1975
47. Herbert, D.: Proc. 6th Intl. Symp. Continuous Culture of Microorganisms, Chichester: Ellis Horwood 1976
48. Hikuma, M. et al.: Biotech. Bioeng. *21*, 1845 (1979)
49. Hills, J. H.: Chem. Eng. J. *12*, 89 (1976)
50. Ho, C. S. et al.: Biotech. Bioeng. *19*, 1503 (1977)
51. Hsu, H. W.: In Biotechnology and Bioengineering Symp. (Scott, ed.) No. *8*, 1, 1978
52. Huffman-Reichenbach, L. M., Harper, W. J.: J. Dairy Sc. *65*, 887 (1982)
53. Inloes, D. S. et al.: Biotech. Bioeng., to be published (1983a)
54. Inloes, D. S. et al.: Applied Env. Microbiol. (1983b)
55. Kim, S. S., Cooney, D. O.: Chem. Eng. Sci. *31*, 289 (1976)
56. Kiparissides, C., MacGregor, M.: Presented at the 182nd ACS Meeting, New York: Amer. Chem. Soc., Aug. 23–28, 1981
57. Klein, J., Vorlop, K.-D.: In: Foundations of Biochemical Engineering (Blanch et al. eds.), Symp. Ser. No. 207, New York: Amer. Chem. Soc. 1983
58. Kleinstreuer, C., Belfort, G.: In: Synthetic Membrane Processes (Belfort, ed.), New York: Academic Press 1984
59. Kleinstreuer, C. et al.: Comp. Biomed. Res. *16*, 29 (1983)
60. Kleinstreuer, C. et al.: Proc. AIChE Annual Mtg., San Francisco, CA, Nov. 1984
61. Kleinstreuer, C., Poweigha, T.: Biotech. Bioeng. *24*, 1941 (1982)
62. Kuraishi, M. et al.: In: Microbial Growth on C1-Compounds, p. 231, Osaka: Soc. Ferment. Technol., 1975
63. Lee, G. K. et al.: Biotech. Bioeng. *23*, 487 (1981)
64. Luttmann, R. et al.: ibid. *24*, 817 (1982); 24, 1851 (1982)
65. Luttmann, R. et al.: Comp. Chem. Eng. *7*, 43 (1983); *7*, 51 (1983)
66. Maiorella, B., Wilke, C. R.: Biotech. Bioeng. *22*, 1749 (1980)
67. Margaritis, A., Sheppard, J. D.: ibid. *23*, 2117 (1981)
68. Margaritis, A., Zajic, J. E.: ibid. *20*, 939 (1978)
69. Mattiasson, B., Ramstorp, M.: Biotechnol. Lett. *3*, 561 (1981)

70. a) Merchuk, J. C., Stein, Y.: Biotech. Bioeng. *23*, 1309 (1981);
b) Merchuk, J. C., Stein, Y.: AIChE J. *27* (1981)
71. Merchuk, J. C. et al.: ibid. *22*, 1189 (1980)
72. Messing, R. A. (Ed.): Immobilized Enzymes for Industrial Reactors, Academic Press 1975
73. Metz, B. et al.: Adv. Biochem. Eng. *11* (1979)
74. Michaels, A. S.: Desalination *35*, 329 (1980)
75. Moo-Young, M., Blanch, H. W.: Adv. Biochem. Eng. *19* (1981)
76. Moo-Young, M., Blanch, H. W.: In: Foundations of Biochemical Engineering (Blanch et al. eds.), Symp. Ser. No. 207, New York: Amer. Chem. Soc. 1983
77. Nauman, E. B., Buffham, B. A.: Mixing in Continuous Flow Systems, New York: Wiley Interscience 1983
78. Orazem, M. E., Erickson, L. E.: Biotech. Bioeng. *21*, 69 (1979)
79. Ovano, A. C.: In: Advances in Chromatography (Giddings et al., eds.), Vol. 15, Marcel Dekker 1979
80. Patterson, G. K.: Chem. Eng. Commun. *8*, 25 (1981)
81. Prokop, A.: Int. J. General Syst. *8*, 7 (1982)
82. Prokop, A.: In: Foundations of Biochemical Engineering (Blanch et al. eds.), Symp. Ser. No. 207, New York: Amer. Chem. Soc. 1983
83. Rajagopalan, R., Tien: AIChE J. *22*, 523 (1976)
84. Ramachandran, P. A.: Biotech. Bioeng. *17*, 211 (1975)
85. Reuss, M., Buchholz, K.: ibid. *21*, 2061 (1979)
86. Rony, P. R.: ibid. *13*, 431 (1971)
87. Roy, T. B. V. et al.: Biotech. Letters *4*, 483 (1982)
88. Russell, T. W. F. et al.: Biotech. Bioeng. *16*, 1261 (1974)
89. Schaftlein, R. W., Russell, T. W. F.: Ind. Eng. Chem. *60*, 12 (1968)
90. Schügerl, K.: Chem. Ing.-Tech., 52, Nr. *12*, s. 951 (1980)
91. Schügerl, K. et al.: Adv. Biochem. Eng. *8* (1978)
92. Schügerl, K.: ibid. *19* (1981)
93. Schügerl, K.: ibid. *22* (1982)
94. Scott, C. D., Hancher, C. W.: Biotech. Bioeng. *18*, 1393 (1976)
95. Shah, Y. T. et al.: AIChE J. *28*, 353 (1982)
96. a) Shuler, M. L. et al.: J. theor. Biol. *35*, 67 (1972);
96. b) Shuler, M. L. et al.: ibid. *41*, 347 (1973)
97. Shuler, M. L., Domach, M. M.: In: Foundations of Biochemical Engineering (Blanch et al. eds.), Symp. Ser. No. 207, New York: Amer. Chem. Soc. 1983
98. Smith, J. M.: Chemical Engineering Kinetics, McGraw-Hill 1981
99. Stephanopoulos, G., San, K.-Y.: Presented at the 182nd ACS Meeting, New York City, Aug. 23–28, 1981
100. Taguchi, H.: Adv. Biochem. Eng. *1* (1971)
101. Tanaka, H. et al.: Biotech. Bioeng. *23*, 1683 (1981)
102. Thoenes, D.: Chem. Eng. Sc. *35*, 1840 (1980)
103a. Tirrell, M., Middleman, S.: Biotech. Bioeng. *17*, 299 (1975)
103b. Trevan, M. D.: Immobilized Enzymes, John Wiley 1980
104. Tsao, G. T., Lee, Y. H.: In: Annual Reports on Fermentation Processes (Pearlman, ed.), Vol. 1, New York: Academic Press 1977
105. U.S. Dept. of Energy: Contact: K. W. Koch, JPL, Pasadena, CA 91109 (1983)
106. Van Keulen, M. A. et al.: Biotech. Bioeng. *23*, 1437 (1981)
107. Vemuri, V.: Modeling of Complex Systems, New York: Academic Press 1978
108. Vemuri, V., Karplus: Digital Computer Treatment of Partial differential Equations, Prentice-Hall 1981
109. Venkatsubramanian, K. (Ed.): Immobilized Microbial Cells, Symp. Series 106, Amer. Chem. Soc. 1979
110. Verhoff, F. H., Schlager, S. T.: Biotech. Bioeng. *23*, 41 (1981)
111. Wang, D. I. C. et al.: Fermentation and Enzyme Technology, John Wiley 1979
112. Wang, H. Y. et al.: Biotech. Bioeng. *21*, 975 (1979)
113. Waterland, L. R. et al.: AIChE J. *20*, 50 (1974)

114. a) Webster, I. A., Shuler, M. L.: Biotech. Bioeng. *20*, 1541 (1978)
114. b) Webster, I. A. et al.: ibid. *23*, 447 (1981)
115. Webster, I. A. et al.: ibid. *21*, 1725 (1979)
116. Yau, W. W. et al.: Modern Size-Exclusion Liquid Chromatography — Practice of Gel Permeation and Gel Filtration Chromatography, John Wiley 1979
117. Zeigler, B. P. et al. (eds.): Methodology in Systems Modeling and Simulation, North-Holland Publishing Company 1979

Fed-batch Techniques in Microbial Processes

Tsuneo Yamanè and Shoichi Shimizu
Laboratory of Bioreaction Engineering, Department of Food Science and Technology, Faculty of Agriculture, Nagoya University, Chikusa-ku, Nagoya 464, JAPAN

This article focuses on the recent world-wide advances of the biotechnology of the increasingly important fed-batch (or semi-batch) cultivation technique used in microbial processes. The history and characteristics of the fed-batch technique in microbial reactions are reviewed and examples of those fed-batch operations are cited which have greatly increased the productivity of microbial conversions in comparison to conventional batch operation; the various fed-batch techniques will be classified according to the mode of nutrient-feeding. Theoretical mathematical models are compared with experimental results. Fed-batch cultures with automatic feedback control are discussed in detail. Cultivation of high microbial cell concentrations, repeated fed-batch operations, and the start-up of microbial processes are described and, finally, some future prospects of fed-batch techniques in microbial processes are discussed.

1 Introduction

In microbial as well as chemical processes, two different modes of operation, batch and continuous, have been applied. In a batch microbial process, all nutrients required during one run of cultivation, except for molecular oxygen in an aerobic process and ammonia or other chemicals for pH adjustment, are added to the medium before cultivation is started, and the final products are only removed at the end of each batch run. In a continuous microbial process, all nutrients are continuously added to the bioreactor and fractions of the medium are removed at the same flow rate as that of supplied nutrients to maintain a constant culture volume. Next to these "ideal" conditions exist many modifications concerning the mode of addition of raw material(s) and withdrawal of products.

In the broadest sense, "fed-batch" is defined as a technique in microbial processes where one or more nutrients are supplied to the bioreactor during cultivation and in which the products remain in the containment until the end of the run. In some cases, all nutrients are gradually fed into the bioreactor. Fig. 1 shows a concept of a fed-batch technique. In reference to the aforementioned definition, it may be regarded as a modification of batch operation. Alternatively, it is sometimes called "semi-batch" in English, "Zulaufverfahren" in German, and "Ryūkaho" in Japanese.

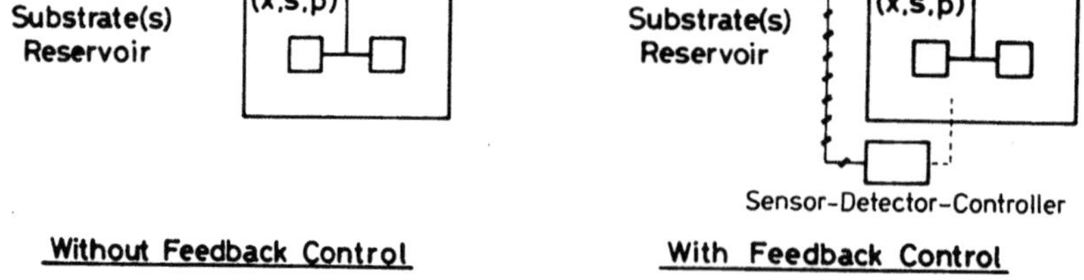

Fig. 1. Schematic diagram of fed-batch operation in a microbial process [11]

A similar type of microbial cultivation was formerly named "extended culture", in which the limiting substrate concentration was controlled by manipulating its feed rate [1]. Although Pirt [2] had claimed that the term "fed-batch culture" was used by us in 1973 to refer to a batch culture continuously fed with nutrient medium [3], we had actually found the term in a review article by Burrows in the book entitled "The Yeast" [4]. The term "semi-batch" operation is commonly used in chemical engineering. Here, the reactor (normally a stirred tank) is first partially filled with reactant, and additional reactants are added successively until the desired end products are formed. Fed-batch, therefore, is clearly identical to semi-batch. Of the various terms, we prefer "fed-batch", as the substrates added to the reactor in microbial processes are nutrients taken up by microorganisms as feedstuff. The term is now often quoted in biochemical engineering and has broadly established.

The basic characteristic of fed-batch microbial processes is that the concentrations of nutrients fed into the culture liquid of the bioreactor can be controlled voluntarily by changing the feed rate. In an "ideal" batch bioreactor, nutrient concentrations are not controlled and are progressively used up; whereas in an "ideal" continuous bioreactor at steady state, the concentrations of all nutrients including the growth-limiting substrate, are kept constant. With respect to the microorganisms environment, fed-batch falls between ideal batch and continuous operations.

2 Utility of the Fed-batch Techniques

Generally speaking, fed-batch is superior to conventional batch operation especially when changing concentrations of a nutrient (or nutrients) affect the yield or productivity of the desired metabolite. There are also several other minor advantages of medium feeding. The types of processes for which fed-batch operation is effective and the corresponding minor advantages can be summarized as follows:

2.1 Substrate Inhibition

Nutrients such as methanol, ethanol, acetic acid, and aromatic compounds inhibit the growth of microorganisms even at relatively low concentrations. By adding such substrates properly lag-time can be shortened and the inhibition of cell growth markedly reduced.

2.2 High Cell Concentrations

To achieve very high cell concentrations, e.g., 100 g of dry cells per l, in a batch culture a high concentration of nutrients in the medium is needed. At such high concentrations nutrients become inhibitory, even though they have no such effect at the normal concentrations used in batch cultures.

2.3 Glucose Effect

In the production of baker's yeast from malt wort or molasses it has been recognized since the early 1900s that ethanol is produced even in the presence of sufficient dissolved oxygen (DO) if an excess of sugar is present in the culture liquid. Ethanol is a main cause of low cell yields. Aerobic ethanol formation in the presence of high glucose concentrations is known as the glucose effect. To reduce this effect, a fed-batch process is generally employed for baker's yeast production.

2.4 Catabolite Repression

When a microorganism is provided with a rapidly metabolizable carbon-energy source such as glucose, the resulting increase of the intracellular concentration of ATP leads to the repression of enzyme biosynthesis, thus causing a slower metabolization of the energy source. This phenomenum is known as catabolite repression. Many enzymes, especially those involved in catabolic pathways, are subject to this repressive regulation. A powerful method of overcoming catabolite repression in enzyme biosynthesis is a fed-batch culture in which the glucose concentration in the culture liquid is kept low, where growth is restricted, and enzyme biosynthesis is derepressed [5-8].

2.5 Auxotrophic Mutants

In a microbial process employing an auxotrophic mutant (nutritional mutant), excess supply of the required nutrient results in abundant cell-growth with little accumulation of the desired metabolite due to feedback inhibition and/or end-product repression. Starvation of the required nutrient, however, lowers cell-growth as well as the overall production of the desired metabolite, as the production rate is usually proportional to the cell concentration. In such a process, the accumulation of the desired metabolite can be maximized by growing the mutant on a limited amount of the required nutrient. To cultivate the mutant on a low concentration of required nutrient, the nutrient is fed to the batch culture at a controlled rate. This technique is commonly used in industrial amino acid production with auxotrophic mutants.

2.6 Extension of Operation Time

A culture can be operated for an extented period of time by adding nutrient(s) at the end of batch operation. If the desired metabolite is still being produced at this time, its yield, i.e. final concentration at the time of harvest, will be increased by extending the culture time. However, an increase of productivity (final concentration divided by the total culture period, i.e., running cost) cannot always be expected as extented cultivating proportionally increases running costs. From the viewpoint of effective utilization of the accumulated biocatalysts, cultivation should be extended if cells still actively excrete the desired metabolite.

In a nongrowth-associated microbial process, such as antibiotics production, microorganisms initially rapidly utilize the carbon-energy source for growth and then synthesize the desired secondary metabolite in the subsequent declining phase and early stationary phase. In a conventional batch process, this production phase is short, due to the depletion of the carbon-energy source; the subsequent cell autolysis is rapid and severe. Therefore, after transition from growth to synthesizing phase, it is important to maintain a concentration of the carbon-energy source where the microorganisms are semi-starved but where enzyme activity for synthesis is highest. To prolong and maintain high synthesis activity of the semi-starved microorganism, the carbon-energy source and/or precursors are fed into the bioreactor at a controlled rate.

2.7 Replacement of Water Lost by Evaporation

In aerobic microbial processes requiring extended reaction periods, such as in antibiotics production (1–2 weeks), considerable amounts of water are lost as the vapor form through the exhaust gas. For example, if the exhaust gas in an aerobic microbial process operated at 30 °C with a 1.0 vvm supply of air (60% relative humidity), is saturated with water, the total water loss after 2 weeks may amount to 25% of the initial culture broth. This leads to a considerable concentration of the mycelial broth and an accompanying change in its rheological behavior. This concentration effect can be counteracted by controlled feeding of nutrients at a rate equivalent to the average rate of water loss by evaporation [9].

2.8 Decreasing Viscosity of Broth

In microbial biopolymer production of, e.g., dextran, pullulan, or xanthan gum, broth viscosity can be kept low by continuous feeding of nutrients. Otherwise, the markedly increased viscosity of the broth will raise the agitation power consumption and lower oxygen transfer efficiency. Blanch et al. [10] suggested that a fed-batch process in which the nutrient was fed shotwise following withdrawal of an equal volume of broth from the vessel, would allow a far greater reduction in viscosity than continuous feeding.

It should be emphasized, however, that in actual microbial processes the feeding of the nutrient(s) solution produces some of these effects inseparably. For example, effect (7) always occurs, even in cases where the broth volume is considerably increased by the feeding. Effects (1) and (2) combine in SCP production from methanol, and effects (1) and (3) are common in baker's yeast production. The combination of effects (6) and (7) is important in antibiotics production.

3 Classification of Fed-batch Techniques

The major purpose of fed-batch operation is to control the substrate (nutrient) concentration within the liquid culture. In establishing a fed-batch technique, two initial decisions must be made, a) on the type of nutrient, and b) on the mode of application. To identify the most effective nutrient(s), knowledge of microbial physiology, biochemistry, and genetics is essential; the possibility of substrate-inhibition must obviously be considered. Sometimes it is inevitable to work by trial-and-error to find an effective substrate. On the other hand, biochemical engineering may play an important role in deciding on the optimal feeding mode.

It is possible to classify fed-batch microbial processes according to the feeding mode. Table 1 shows the classification by Yamanè [11]. The author classifies fed-batch operations broadly into: those processes without feedback control, and those with feedback control. These are further subdivided with reference to feed rates, into: constant, exponential, and optimized fed-batch operations. In the fed-

Table 1. Classification of fed-batch operations in microbial processes [11]

1. Without feedback control
   - 1.1 Intermittent addition
   - 1.2 Constant rate
   - 1.3 Exponentially increased rate
   - 1.4 Optimized
   - 1.5 Others
2. With feedback control
   - 2.1 Indirect feedback control
   - 2.2 Direct feedback control
   - 2a Constant-value control
   - 2b Optimal control

batch operation with direct feedback control, the concentration of the substrate in the culture liquid directly controls the addition rate of further substrate, whereas in the fed-batch operation with indirect feedback control, other feedback parameters such as DO, pH, or RQ are used. In both cases, the substrate concentration in the culture is either kept at a constant level (constant-value control) or changed optimally (optimal control). Initially, rather simple feeding modes such as intermittent addition or constantly-fed-batch operations were commonly used. Recent technical advances for measurement and control in biotechnology have made more sophisticated feeding modes possible.

Table 2.1. Fed-batch operations for baker's yeast production

Fed substrate(s)	Feedback parameter	Feeding mode	Ref.
		1. Without feedback control	
Molasses	—	Intermittent and incremental addition	14)
		2. With feedback control	
Molasses	DO	On-off control	15)
Molasses	DO	On-off control	16)
Wort	DO	On(50s)-off(140s) control	17)
Semisynthetic medium or molasses	RQ	Feed rate is increased stepwise manually	18)
Molasses	RQ	On-off control	19)
Molasses	RQ	Combined anticipatory/feedback control aided by computer	13)
Glucose	Dissolved ethanol	On-off or PID control	20)
Glucose	Dissolved ethanol	Advanced control system	21)
Semisynthetic medium	Ethanol vapor in exit gas	Stepwise change	22)
Synthetic medium	Ethanol in exit gas	Feedforward and feedback controls aided by microcomputer	23)

4 Survey of Fed-batch Microbial Processes

Fed-batch processes have been performed in the production of a wide variety of products and with the perspective to increase the yields of desired metabolites. Products include baker's yeast, SCP (microbial cell mass), antibiotics, vitamins, enzymes, amino acids and others. A literature survey of these are summarized in Tables 2.1–2.5. Details of industrial fed-batch processes are rarely revealed, since they are know-hows of each private firm. Little information is available on the actual increases in yield or productivity offered by the fed-batch operation in comparison with the conventional batch operation, and hence these data are omitted from the Tables.

Table 2.2. Microbial cell mass production by fed-batch culture

Microorganism	Fed substrate	Feedback parameter	Feeding mode	Ref.
		1. Without feedback control		
Pseudomonas sp.	Methanol	—	Constant rate	24)
a bacterium	Methanol	—	Exponentially increased rate	25)
Cellulomonas sp.	Cellulose	—	Constant rate but conc. of the nutrients increased linearly	26)
Pichia methanothermo	Methanol	— —	Constant rate as vapor or intermittent/incremental addition	27)
Candida utilis	Ethanol	—	Stepwise according to addition schedule	28)
		2. With Feedback control		
Candida utilis	Acetic acid	pH	On-off (?), automatic	1)
Candida boidinii	Methanol	Methanol vapor in exit gas	On-off, automatic	29)
Pseudomonas AM-1	Methanol-ammonia solution	pH	On-off, automatic	30)
Protaminobacter ruber	Methanol	DO	Intermittent, manual	31)
Protaminobacter ruber	Methanol	Methanol	Intermittent, manual	32)
Candida brassicae	Ethanol	Ethanol	Intermittent, manual	32)
Candida brassicae	Ethanol	DO	Intermittent, manual	33)
Candida utilis	Ethanol	Ethanol vapor in exit gas	On-off, automatic	34)
Pichia farinosa	Ethanol	Ethanol	Continuous addition, with control parameter measured by porous Teflon tubing sensor	35)
Candida brassicae	Ethanol	Ethanol	Automatic, by a porous Teflon membrane-covered ethanol probe	36)

4.1 Baker's Yeast

There is a long history of baker's yeast production by means of fed-batch processes. Much progress has been made on feeding modes, as shown in Table 2.1. Fed-batch operation is generally employed to circumvent the repressive effect of free glucose

Table 2.3. Production of antibiotics by fed-batch operation

Fed substrate	Feeding mode	Ref.
1. Penicillin production by *Penicillium chrysogenum*		
Nutrients	Intermittent	37)
Phenylacetic acid	Intermittent	38)
Phenylacetic acid	Intermittent	39)
Lactose	Intermittent	40)
Glucose or sucrose	Intermittent or constant feed rate	41)
Phenylacetic acid	Intermittent	42)
Glucose	Constant feed rate	43)
Glucose, lactose or sucrose	Constant feed rate	44)
Glucose	Constant feed rate	45)
Phenylacetic acid and lard oil	Intermittent	46)
Glucose and soybean oil	Intermittent	47)
Phenylacetic acid and N-source	Intermittent	47)
Nutrients	Predetermined pattern and constant feed rate afterward	48)
Phenylacetic acid	Controlled release from tablets	49)
Sucrose, phenylacetic acid, and NH_4NO_3	Intermittent	50)
Sugar and phenylacetic acid	Sugar addition automatic feedback control by pH and phenylacetic acid added at constant feed rate	51)
Glucose	Automatic controlled by DO as feedback parameter	52)
Sugar	Continuous, constant specific feed rate [g (g cell)$^{-1}$ h^{-1}]	53)
Carbohydrate	Constant feed rate or optimized feed rate	54)
Sucrose	Varied as to keep the concentration constant (theoretical study)	55)
Glucose, NH_3, and phenylacetic acid	Glucose addition controlled through computer-aided on-line mass balancing	56)
2. Other antibiotics		
*Streptomycin production by *Streptomyces griseus*		
Glucose and $(NH_4)_2SO_4$	Intermittent	57)
*Novobiocin production by *Streptomyces niveus*		
Nutrients	Intermittent	58)
*Griseofulvin production by *Penicillium griseofulvum*		
Sugar	Feed rate changed with time, but predetermined pattern	59)
*Candidin and Candihexin production by *Streptomyces viridoflavus*		
Glucose	Intermittent	60)
*Cephalosporin C production by *Cephalosporium acremonium*		
Glucose and methionine	Constantly fed	61)

Table 2.4. Amino acid production by fed-batch microbial processes

Amino acid	Fed substrate	Feeding mode	Ref.
Glutamic acid	Urea	Intermittent	62)
	Urea	Intermittent	63)
	Acetic acid	Intermittent	64)
	Ethanol and urea	Intermittent	65)
	Ammonium benzoate	Intermittent	66)
	Ethanol	Optimized, on-off control	67)
Tyrosine	Phenol	Intermittent	68)
Lysine	Nutrient (defatted soybean hydrolyzate	Optimized	69)
Tryptophan	Athranilic acid	Constant feed rate	70, 71)

Table 2.5. Fed-batch experiments for enzyme formation

Enzymes	Microorganism	Fed substrate	Feeding mode	Ref.
β-Galactosidase	*Escherichia coli*	Nutrient	Constant feed rate	5)
Polygalacturonic acid trans-eliminase	*Aeromonas liquefaciens*	Carbon source	Constant feed rate	6)
Cellulase	*Myrothecium varrucaria*	Glucose	Constant feed rate	7)
Cellulase	*Pseudomonas fluorescens var. cellulosa*	Carbon source	Constant feed rate	8)
Isoamylase	*Aerobacter aerogenes*	Soluble starch	Constant feed rate	72)
β-Galactosidase	*Escherichia coli*	Glycerol	Constant feed rate	73)
β-Glucanase	*Bacillus subtilis*	Starch	Constant feed rate	74)
β-Amylase	*Bacillus megaterium*	Carbon source	Constant and varied feed rate	75)
Cellulase	*Trichoderma reesei*	Cellobiose or cellulose and cellobiose	Addition controlled by CO_2 gas pressure in the exhaust gas as feedback parameter	76)
Cellulase	*Trichoderma reesei*	Soluble carbon	Addition controlled by CO_2 gas pressure in the exhaust gas as feedback parameter	[illegible])
Cellulase	*Trichoderma viride*	Cellulose and ammonia	Intermittent addition of cellulose and automatic feeding of ammonia by pH as feedback parameter	78)

("glucose effect") mentioned in 2.3. The critical concentration at which glucose tends to be partly metabolized to ethanol has been measured by several researchers. The observed values show great variation (3.6–153 mg l^{-1}) which might be due to differences in strains or physiological conditions [12]. Wang et al. [13] reported 130 mg l^{-1} of glucose using a mixture of equal volumes of cane molasses and beet molasses. Since the fed raw material (carbon-energy source) in industry is beet or cane molasses, it is difficult to directly measure the concentration of the reducing sugars in the culture liquid. Thus, feeding of the nutrient is indirectly feedback controlled by use of one of the feedback parameters in Table 2.1.

4.2 SCP (Microbial Protein)

SCP (microbial protein) production is expected to play a significant role in efforts to solve worldwide protein shortages. It has a considerable potential as a cheap and rapidly manufacturable protein feed for farm animals, as well as directly as nutrients for humans. Although long-term continuous culture of microorganisms may be the most economic process for SCP production, a number of papers have reported fed-batch cultures on methanol, ethanol, or acetic acid, all of which show severe substrate-inhibition of microbial cell growth. Table 2.2 summarizes several of these reports. For volatile carbon-energy sources, like methanol or ethanol, automatic feedback control of the feeding is possible by using feedback parameters which maintain optimal concentrations in the culture broth.

4.3 Antibiotics

A number of studies have been carried out on the effect of nutrient feeding in antibiotics production, particularly fed-batch studies to improve penicillin titers.

4.3.1 Penicillin

In penicillin production, both the main energy source (e.g., glucose, sucrose, molasses) and the precursor of the penicillin molecule (e.g., phenylacetic acid) are added during cultivation. Expensive lactose, originally used as a slowly assimilable carbon source, was eventually replaced by rapidly assimilable carbohydrates, such as glucose, sucrose, or molasses added by a slow feeding method. Although the initial addition of phenylacetic acid, a precursor for the biosynthesis of penicillin G, was recognized to considerably increase production, the amount incorporated by the mycelium was very low as most of it was catabolized, and high concentrations of the compound proved to inhibit the growth of the microorganism. Continuous feeding techniques have helped to overcome both of these problems. It was also found that dropwise (intermittent) addition of an oily substance, such as lard oil, as an antifoam agent increased the yield of penicillin.

Table 2.3 presents a literature survey of penicillin production by fed-batch operation. Intermittent addition or continuous supply with constant feed rates have been the major feeding modes reported, but recently several researchers have studied time-optimal feed rates by applying both system kinetics and rigorous optimization theory. Except for two papers dealing respectively with indirect control by use of pH and of DO as a feedback parameter, however, feedings have been done without feedback control, probably because of difficulty in finding a suitable feedback parameter.

4.3.2 Other Antibiotics

Few papers make notice of nutrient feeding in the production of other antibiotics. Works available to us are listed in Table 2.3.

4.4 Amino Acids

In some techniques for amino acid production, carbon sources, nitrogen sources, and surface active agents can be fed to obtain increased yields (Table 2.4). In this field most work has been done in Japan. Nutrients are fed not only to increase product yield but in some cases also to control the pH within a suitable range.

4.5 Enzymes

Various enzymes are being industrially produced by microbial processes. Astonishingly, only limited work has been published on fed-batch experiments, probably because existing methods in industrial production are guarded as private know-how. Some papers published by academic researchers are summarized in Table 2.5. The increase in enzyme formation afforded by substrate-feeding is mostly due to derepression of enzyme biosynthesis by keeping substrate concentrations extremely low.

4.6 Other Metabolites

The yield of riboflavin by *Ashbya gossypii* has been increased by intermittent feeding of both glucose and inositol at the same time [79]. This method was attempted because high glucose concentrations inhibited vitamin production and high inositol concentrations lowered cellular synthesis activity, even though inositol was the growth factor of the strain. By semi-batch culture of methanol-assimilating bacteria with exponentially increased methanol feed, vitamin B_{12} formation could be raised to 2.46 mg l^{-1} [25]. A variety of patented fed-batch techniques have been used to improve yields of vitamin B_{12} by different microorganisms [80]. In the production of gibberellic acid by *Fusarium moniliforme*, slow feeding at a constant rate starting 24 h after inoculation eliminated the need to use such slowly assimilated carbon sources as glycerol or lactose [81]. A high final concentration of acetic acid in vinegar, e.g., exceeding 17% acidity, has been obtained by an increase in the feeding of ethanol depending on cell growth starting from a low ethanol concentration [82]. This technique has been adopted mainly to avoid substrate-inhibition and also to allow the microorganisms to acclimate gradually to a high concentration of acetic acid. *Gluconobacter suboxydans* was cultivated in a fed-batch culture with a DO-stat, resulting in 628 g l^{-1} of sorbose from sorbitol after 14 h [83]. This yield is surprisingly high compared to previously reported figures. *Lipomyces starkeyi* was found to produce a large amount of neutral lipids by microcomputer-aided fed-batch process [84]. By reducing the concentration of ammonium ion in the broth after 48 h of fed-batch culture, high yields of biomass (153 g l^{-1}) and lipids (83 g l^{-1}) were obtained after 140 h. It has been suggested that in fed-batch alcohol fermentations, concentrated glucose should be added to the culture medium, so that glucose concentration may not exceed 0.2 g l^{-1} [85]. Intervals in glucose addition were expressed as a function of cell mass concentration during operation.

5 Mathematical Theory of Fed-batch Cultivations without Feedback Control

In fed-batch procedures without feedback control, the feed rate is changed during operation according to a predetermined pattern. Even if the nutrient(s) is supplied to the bioreactor without feedback control, the time course of the culture is not significantly different than expected and is reproducible provided that the physiological state of the inoculum and initial cultivation conditions are prescribed.

In the most primitive fed-batch culture technique without feedback control listed in Table 1, the substrate is intermittently added. This procedure will not be further described in this review although a number of papers have appeared on this mode of addition in laboratory-scale experiments, most likely as of easy handling. It is obvious that frequent intermittent addition approaches continuous feeding. We can find no particular advantages in the use of intermittent feeding.

5.1 Mathematical Models

Assume a well mixed bioreactor in which the growth of microorganisms is limited by the concentration of a growth-limiting substrate, where all other nutrients are present in excess. Only the growth-limiting substrate or its aqueous solution containing all other necessary nutrients is supplied to the bioreactor at a constant or varied feed rate, and no culture broth is withdrawn from the vessel. Feeding of the substrate is started only after a lag-phase when the microorganisms grow actively. If the growth-limiting substrate is an energy source, part of it will be used for maintenance of biomass.

With these assumptions, the simplest mathematical model for the fed-batch culture is as follows:

$$\frac{d(vx)}{dt} = \mu vx \tag{1}$$

$$\mu = g(s) \tag{2}$$

$$\frac{d(vs)}{dt} = f_v s_{in} - \frac{1}{k_y}\frac{d(vx)}{dt} - mvx \tag{3}$$

$$\frac{dv}{dt} = f_v + f_{acid} + f_{base} + f_{antifoam} - f_{vap} \tag{4}$$

$$q_p = h(\mu) = h\{g(s)\} \tag{5}$$

In these equations μ and q_p correspond to:

$$\mu \equiv \frac{1}{vx}\frac{d(vx)}{dt} \tag{6}$$

$$q_p \equiv \frac{1}{vx}\frac{d(vp)}{dt} \tag{7}$$

Equation (1) describes the growth of the microorganism, in which μ is assumed to be a function of the concentration of the fed substrate in the culture liquid, according to Eq. (2). For variable-volume fed-batch cultures, μ is defined by Eq. (6). Equation (3) formulates the balance of the fed growth-limiting substrate in the culture liquid. The first term on the right-hand side of the equation represents substrate input, the second term is the consumption rate of the substrate calculated from the cell growth rate, and the last term is the substrate consumption rate for maintenance of cellular metabolism. If the substrate is volatile and the culture is aerobic, the loss of substrate by aeration should be considered. When the balance of an element constituting the fed-substrate is formulated, the elemental composition of the dry cell mass and of all metabolites is required. Equation (4) shows that the increase in the culture broth results from the volumetric feed rate of the substrate solution, from the addition rate of acid/base solutions for pH control, from the feed rate of antifoam agent solution, and from the evaporation rate of water in the exhaust gas. The latter can be neglected for short-term operations but not for long-term aerobic processes as in antibiotics production [9]. It is clear that $k_{vap} = 0$ when the culture is anaerobic. In Eq. (5) the specific production rate q_p is a function of μ, this being often adopted in microbial reactions. Since μ is a function of s, q_p becomes a function of s as well. Equation (5) indicates that q_p can be regulated through control of s.

If the following can be assumed: $(f_{acid} + f_{base} + f_{antifoam} - f_{vap}) \ll f_v$, then Eq. (4) becomes

$$\frac{dv}{dt} = f_v \tag{8}$$

Substitution of Eq. (8) into Eq. (1) gives

$$\frac{dx}{dt} = (\mu - f_v/v)\,x \tag{9}$$

Following the proposal by Pirt [86], we introduce the dilution rate D as:

$$D \equiv \frac{f_v}{v} \tag{10}$$

Then, Eqs. (9) and (3) become:

$$\frac{dx}{dt} = (\mu - D)\,x \tag{11}$$

$$\frac{ds}{dt} = D(s_{in} - s) - \left(\frac{\mu}{k_y} + m\right) x \tag{12}$$

Equations (11) and (12) are formally identical with the corresponding balance equations in a chemostat (continuous culture) [89]. It should be noted, however, that D in Eq. (10) is time-variable and not constant, as v increases with time, whereas D of a chemostat is constant. Therefore, the concept of the dilution rate in the kinetics of the fed-batch operation is less useful than in a chemostat. Equations (1) to (12) provide a model of a culture in which the culture volume is variable (increased).

In some microbial processes the culture volume may be virtually constant when the fed substrate is an undiluted liquid, a very concentrated solution, a gas, or solid. If the culture volume is constant, Eq. (1) is reduced to:

$$\frac{dx}{dt} = \mu x \tag{13}$$

By introducing the mass feed rate f_s defined as:

$$f_v s_{in} \equiv f_s \tag{14}$$

Equation (3) is simplified to:

$$\frac{ds}{dt} = \frac{f_s}{v} - \frac{1}{k_y}\frac{dx}{dt} - mx \tag{15}$$

Now, Eq. (4) becomes unnecessary. Eventually, the mathematical model for constant volumes becomes much simpler than that for variable volumes.

5.2 Constantly-fed-batch Cultures

A fed-batch culture with a constant feed rate of a growth-limiting substrate (f_v = const.) is called a constantly-fed-batch culture and is regarded as the simplest fed-batch culture system [88]. As we have seen in Sect. 4, most of the earlier fed-batch operations made use of this feeding mode. Mathematical analysis of this type of operation was carried out by Yamanè et al. [3,88], Pirt [86,89], and Dunn et al. [87,90] and experimental studies were by Yamanè et al. [91]. The model of Yamanè et al. assumes Eq. (8) to hold true. The equation is valid for low substrate concentrations in the feed; the increase of the culture volume is mainly attributed to the water which is the major part of the feed. However, the substrate itself is partly converted to new biomass and partly to water, CO_2, and metabolites. Therefore, we assume that the increase of the culture volume arises from the water content of the feed and, to a certain extent from the substrate itself. When the substrate is liquid, the rate of the volumetric increase is:

$$\frac{dv}{dt} = 10^{-3}\left(\frac{f_w}{\varrho_w} + c\,\frac{f_s}{\varrho_s}\right) = f_v \tag{16}$$

where c is the increment of the culture volume per unit volumetric feed rate of pure substrate. It is possible to determine c experimentally by semi-batch culturing with a constant feed of pure substrate. If the substrate is solid, c must be defined on the basis of its mass feed f_s. Assuming that the substrate content in the feed solution is an operating variable and by defining the rate of volume increase through Eq. (16), we can evaluate the effect of dilution without considering substrate feeding.

We adopt the Monod equation for g(s) in Eq. (2):

$$\mu = \frac{\mu_{max} s}{k_s + s} \tag{17a}$$

For simplicity, we assume that no metabolites other than regular biomass is synthesized. In this case the model is constituted by Eqs. (1), (3), (6), (16) and (17a). To simplify the equation and to present the results more generally, the equations are written in dimensionless forms. The dimensionless variables and parameters are defined as:

$$\left.\begin{aligned}
&X = \frac{x}{x_0 + k_y s_0}, \quad S = \frac{s}{s_0 + x_0/k_y} \\
&\Theta = \mu_{max} t, \quad K_s = \frac{k_s}{s_0 + x_0/k_y} \\
&F_s = \frac{f_s}{(s_0 + x_0/k_y)\, v_0 \mu_m}, \quad F_w = \frac{10^{-3} f_w}{\varrho_w v_0 \mu_{max}} \\
&C = \frac{10^{-3} \varrho_w}{\varrho_s}\left(s_0 + \frac{x_0}{k_y}\right), \quad M = \frac{m k_y}{\mu_{max}} \\
&V = v/v_0, \quad F_v = F_w + C F_s \\
&\hat{\mu} = \mu/\mu_{max}
\end{aligned}\right\} \tag{17b}$$

The nondimensional system equations derived by use of Eq. (17b) were numerically computed for various initial operating conditions and system parameters [87].

It was found that the computed results could be summarized most clearly by the relationship between the total cell mass VX and Θ (Fig. 2). This figure shows that the constantly-fed batch culture is characterized by (arithmetically) linear growth:

$$VX = F_s \Theta + 1 \tag{18a}$$

or in the dimentional form:

$$vx = k_v\{f_s t + v_0(s_0 + x_0/k_v)\} \tag{18b}$$

When $S_0 > K_s$ or $F_s > 1.0$, however, exponential growth appears. As Fig. 3 indicates, the transition from exponential growth to linear growth is very sudden, μ being very low with little change during linear growth. When $F_s \leqq 1.0$ and $S_0 \ll K_s$, exponential growth does not occur, the growth is linear from the beginning of feeding. The dimensionless cell concentration X decreases or increases with increasing Θ depending on the degree of dilution and initial conditions (Fig. 4). X remains constant, however, at $X = 1.0$ almost during the entire course of cultivation when $X_0 = 1$ and $F_s = F_v$. The variations of computed S with Θ corresponding to Figs. 3 and 4

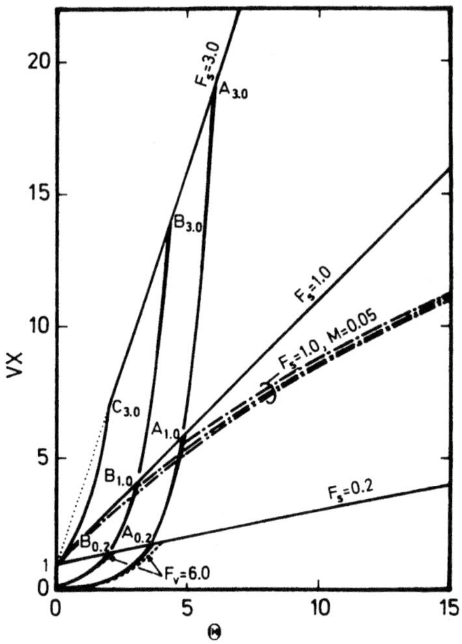

Fig. 2. Total cell mass, VX, as a function of time, Θ for $K_s = 0.005$ and $F_s = 3.0$, 1.0, or 0.2 in the constantly-fed-batch culture [88]. ————, $F_v = 0.0 \sim 2F_s$ and $M = 0.0$; – – – – –, $F_s = 0.2$, $F_v = 6.0$ and $M = 0.0$; —.—.—, $F_s = 1.0$, $F_v = 0.0 \sim 2.0$ and $M = 0.05$. The symbols A, B, and C show transition points from exponential growth to linear growth

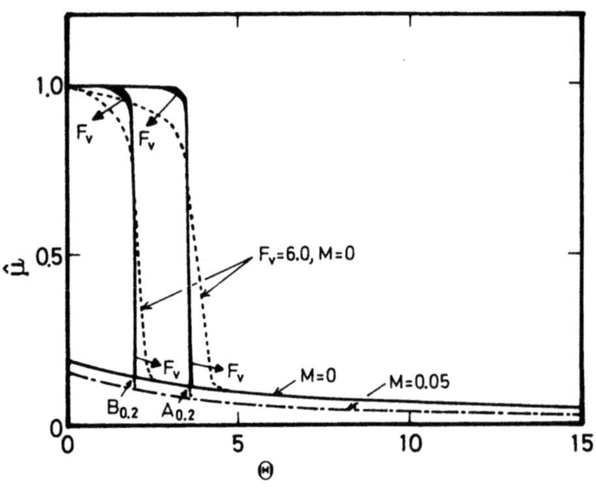

Fig. 3. Variation of relative specific growth rate $\hat{\mu}$ with relative time Θ for $K_s = 0.005$ and $F_s = 0.2$ in the constantly-fed-batch culture [88]. ————, $F_v = 0.0$, 0.1, 0.2, or 0.4 and $M = 0.0$; – – – – –, $F_v = 6.0$ and $M = 0.0$; —.—.—, $F_v = 0.0$, 0.1, 0.2, or 0.4, $M = 0.05$ and $X_0 = 1.0$. Values of $\hat{\mu}$ in the exponential growth phase for $F_v = 0.0$, 0.1, 0.2, or 0.4, $M = 0.05$, and $X_0 = 0.05$ or 0.2 are almost the same as those for $M = 0.0$ and the corresponding values of F_v and X_0

are shown in Fig. 5. These figure indicate that: (a) S decreases sharply at the end of the exponential growth phase, (b) in the linear growth phase, S is lower than K_s and the rate of change of S is quite small, and (c) the value of S during linear growth is independent of F_v. The time courses of S for linear growth with various F_s as para-

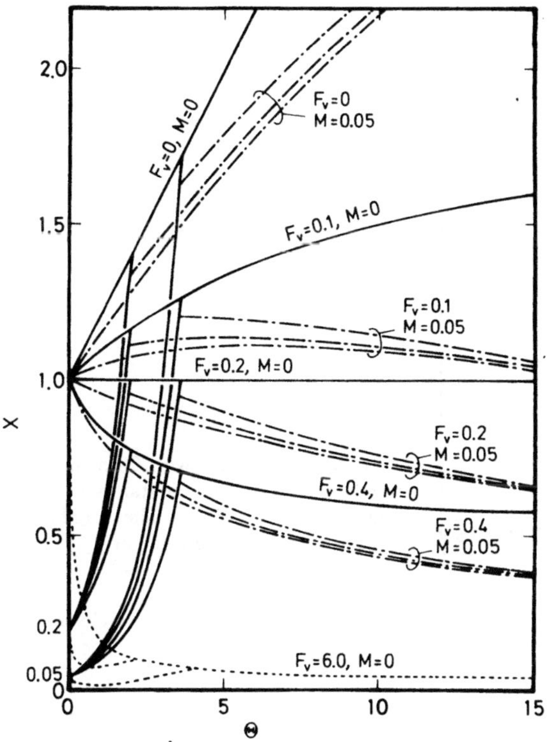

Fig. 4. Time-course of relative cell concentration, X for $K_s = 0.05$ and $F_s = 0.2$ in the constantly-fed-batch culture. The values of X in the exponential growth phase for M = 0.05 agree with those for M = 0.0 [88)]

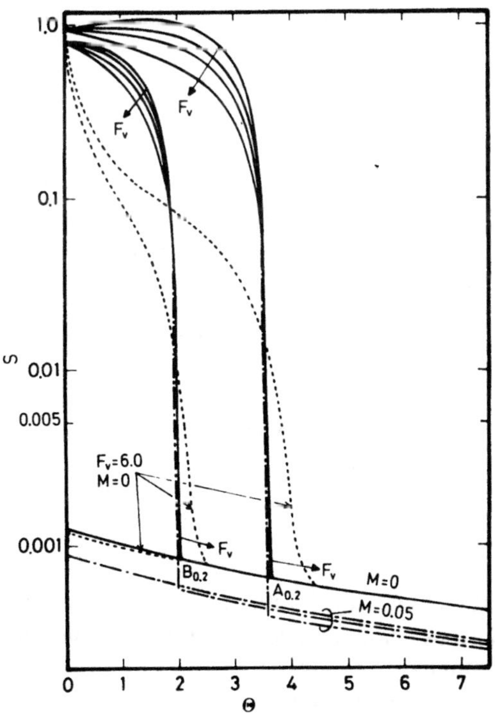

Fig. 5. S vs. Θ for $K_s = 0.005$, $F_s = 0.2$, $F_v = 0.0, 0.1, 0.2, 0.4$, or 0.6 and M = 0.0 or 0.05 in constantly-fed-batch culture [88)]. The number on each line represents the value of F_v. ------, $K_s = 0.005$, $F_s = 0.2$, $F_v = 6.0$, and M = 0.0; —.—.—, $K_s = 0.005$, $F_s = 0.2$, and M = 0.05. The values of S in the exponential growth phase for M = 0.05 are slightly different from those for M = 0.0, but are omitted here

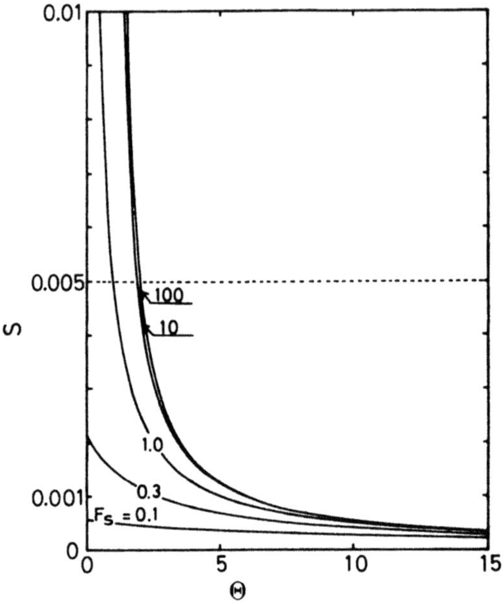

Fig. 6. Variations in S with Θ in the linear growth phase of the constantly-fed-batch culture (K_s = 0.05)[11]. The number on each line represents the value of F_s

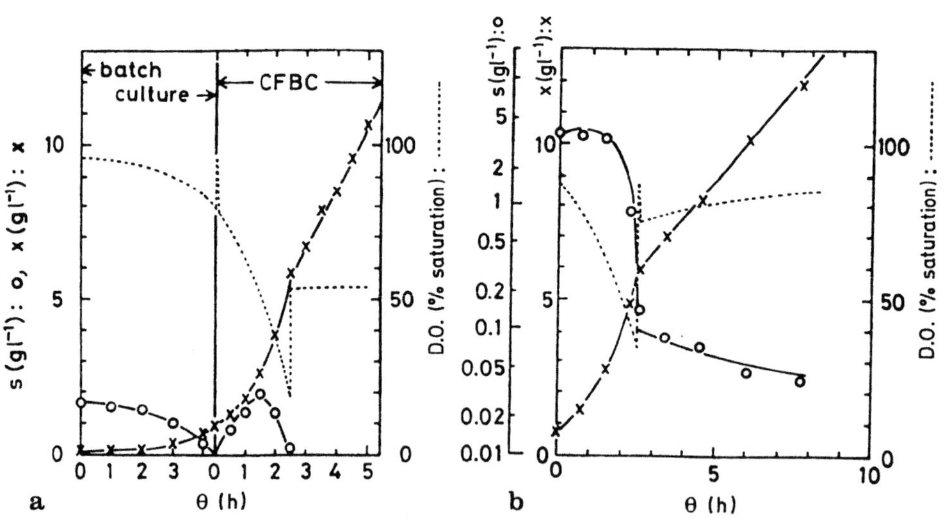

Fig. 7a and b. Time courses of constantly-fed-batch culture with undiluted glycerol feed[91]. Initial glycerol concentrations were (**a**) nearly zero (f_s = 9.20 g h^{-1}, f_w = 0 and v_0 = 2.84 l) and (**b**) 4.0 g l^{-1} (f_s = 5.38 g h^{-1}, f_w = 0 and v_0 = 2.90 l). The solid lines are theoretical curves given by the equations in Table 3

meters are shown in Fig. 6, revealing little change of S with variation of F_s. From these observations, the approximate equations characteristics of constantly-fed-batch cultures have been derived; results are summarized in Table 3.

In order to experimentally verify the approximate equations and to examine the

Table 3. The equation characteristic of constantly-fed-batch cultures [91]

a) Conditions for attaining linear growth from the start of substrate-feeding:

$$s_0 \approx 0$$
$$f_s \leqq x_0 v_0 \mu_{max}/k_y \quad (1)$$

b) x and s in the exponential growth phase

$$x = \frac{v_0 x_0 \exp(\mu_{max} t)}{10^{-3}(f_w/\varrho_w + cf_s/\varrho_s)\, t + v_0} \quad (2)^a$$

$$s = \frac{f_s t + v_0(s_0 + x_0/k_y) - (v_0 x_0/k_y) \exp(\mu_{max} t)}{10^{-3}(f_w/\varrho_w + cf_s/\varrho_s)\, t + v_0} \quad (3)^a$$

c) x and s in the linear growth phase (the succeeding growth phase for m = 0):

$$x = \frac{k_y\{f_s t + v_0(s_0 + x_0/k_y)\}}{10^{-3}(f_w/\varrho_w + cf_s/\varrho_s)\, t + v_0} \quad (4)^a$$

$$s = \frac{k_s f_s}{(\mu_{max} t - 1)\, f_s + \mu_{max} v_0 (s_0 + x_0/k_y)} \quad (5)$$

d) x and s in the succeeding growth phase for m $\neq$ 0:

$$x = \frac{f_s/m - (f_s m - v_T x_T) \exp\{-mk_y(t - t_T)\}}{10^{-3}(f_w/\varrho_w + cf_s/\varrho_s) + v_0} \quad (6)^a$$

$$s = \frac{k_s m k_y (f_s/m - v_T x_T) \exp\{-mk_y(t - t_T)\}}{\mu_{max} t_s/m - (f_s m - v_T x_T)(mk_y + \mu_{max}) \exp\{-mk_y(t - t_T)\}} \quad (7)$$

e) Transition time, t_T:

$$v_0 x_0 \exp(\mu_{max} t_T) = k_y f_s t_T + v_0(x_0 + s_0 k_y) \quad (8)$$

[a] In these equations, $f_v t + v_0$ can be substituted for $10^{-3}(f_w/\varrho_w + cf_s/\varrho_s)\, t + v_0$ with good approximation

characteristics of the culture, *Enterobacter cloacae* NCIB 8271 was cultivated using glycerol as a feed substrate. Figures 7a and 7b present the time courses for the feeding of undiluted glycerol, and show that the experimental data agreed well with the approximate equations. DO rose sharply at the end of the exponential growth phase and maintained an almost constant level during linear growth. This phenomenon was observed throughout and was useful for monitoring the constantly-fed-batch culture. The time course for cultivation with glycerol solutions are shown in Figs. 8a and 8b. When a diluted solution is fed at a high flow rate, x decreases even during exponential growth. In Fig. 9, x was kept constant. Many culture patterns are possible all of which can be well represented by the equations and inequality in Table 3. Table 3 is, hence, useful for the design and operation of a constantly-fed-batch culture.

When v ≈ const. and m = 0, then x, μ, and s during the linear growth phase are given by:

$$\left.\begin{aligned} x &= k_y f_s t + x_0 \\ \mu &= \frac{k_y f_s}{k_y f_s t + x_0} \\ s &\approx s_0 \end{aligned}\right\} \tag{19}$$

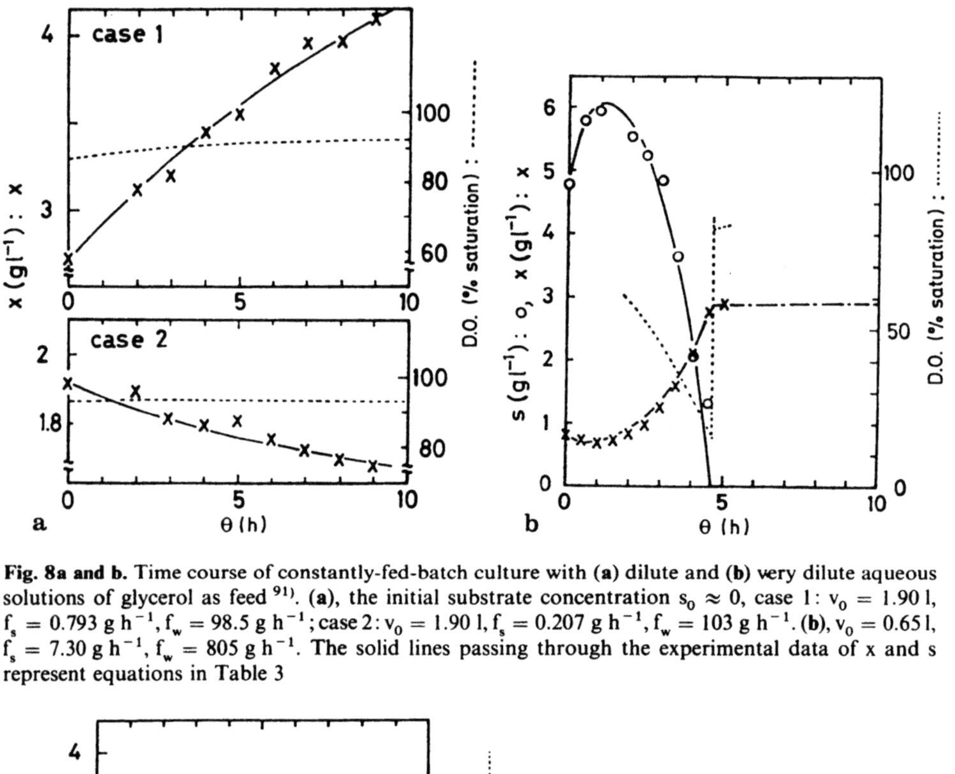

Fig. 8a and b. Time course of constantly-fed-batch culture with (**a**) dilute and (**b**) very dilute aqueous solutions of glycerol as feed [91]. (**a**), the initial substrate concentration $s_0 \approx 0$, case 1: $v_0 = 1.90$ l, $f_s = 0.793$ g h^{-1}, $f_w = 98.5$ g h^{-1}; case 2: $v_0 = 1.90$ l, $f_s = 0.207$ g h^{-1}, $f_w = 103$ g h^{-1}. (**b**), $v_0 = 0.65$ l, $f_s = 7.30$ g h^{-1}, $f_w = 805$ g h^{-1}. The solid lines passing through the experimental data of x and s represent equations in Table 3

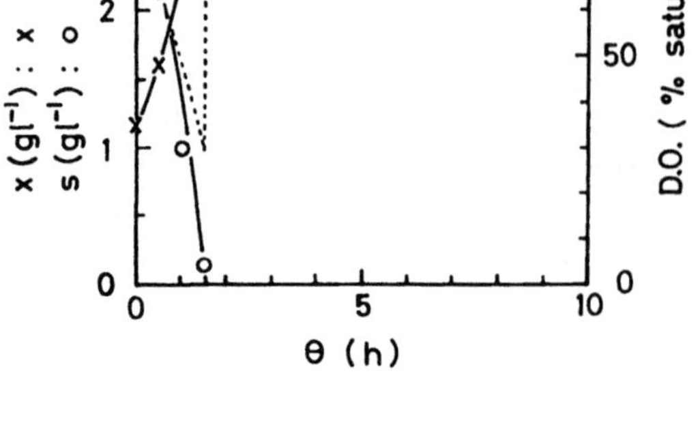

Fig. 9. Time course of a constantly-fed-batch culture in which a "quasi-steady state" was attained ($f_s = 0.367$ g h^{-1}, $f_w = 91.5$ g h^{-1} and $v_0 = 1.86$ l) [91]. The solid lines passing through the experimental data of x and s represent the equations in Table 3

where s_0 must be low enough to assure linear growth. The plot of dx/dt vs. f_s gives a straight line, the slope of which yields k_y. This procedure to determine the average value of k_y was applied by Yamanè et al. [3,91]. The assumption from which Eq. (19) is derived was validated, though Lim [92] had questioned the validity of this assumption.

Linear growth for constant flow rates of nutrients was reported in the early literature [93,94].

5.3 Quasi-steady State in Constantly-fed-batch Cultures

For simplicity, we assume in this section that f_v can be substituted for 10^{-3} $(f_w/\varrho_w + cf_s/\varrho_s)$ and that $m = 0$ in the equations of Table 3. Equation (4) in Table 3 indicates that x increases if $s_{in} > (s_0 + x_0/k_y)$ and decreases if $s_{in} < (s_0 + x_0/k_y)$ during linear growth, and x approaches $k_y s_{in}$ at $t \to \infty$. In between, there is the situation where x remains constant if $s_{in} = s_0 + x_0/k_y$. Therefore, if a fed-batch culture is carried out under the following initial conditions:

$$s_0 < k_s \qquad \text{and} \qquad s_{in} = x_0/k_y \tag{20}$$

then this will subsequently lead to $x = x_0$ (= const.) during the entire linear growth phase. In this event $dx/dt = 0$ so that Eq. (21) can be obtained from Eq. (11):

$$\mu = D = \frac{f_v}{f_v t + v_0} \tag{21}$$

From Eqs. (12) where $m = 0$, (20), and (21), we obtain:

$$s = \frac{v_0}{f_v t + v_0} s_0 \tag{22}$$

Equation (22) states that s is not constant but decreases in time. Pirt named the state of a constantly-fed-batch culture in which $dx/dt \approx 0$, $ds/dt \approx 0$, and $\mu \approx D$ the "quasi-steady state" [86]. This is shown in Fig. 9. It is noteworthy, however, that the condition $ds/dt \approx 0$ is not realized in any linear growth phase as shown in Eq. (22) although $s < k_s$ and ds/dt is small in that phase. The state where $ds/dt = 0$ is only realized in an exponentially-fed-batch culture as will be discussed in detail in Sect. 5.5. Unless the conditions of Eq. (20) are satisfied, the approach towards the "quasi-steady state" is very slow, requiring considerable volume expansions which would prevent observation of the quasi-steady state. Kalogerakis et al. [95] developed a computer-aided control system to force the system rapidly to the quasi-steady state.

When nutrients other than the growth-limiting substrate are fed to the bioreactor together with the substrate, the balance of one of these will lead to the same formula as Eq. (12):

$$\frac{ds^i}{dt} = D(s^i_{in} - s^i) - \left(\frac{\mu}{k^i_y} + m^i\right) x \tag{23}$$

If we assume that $m = m^i = 0$, $s_{in}^i \gg s^i$, and $s_{in} = x_0/k_y$, then:

$$\frac{ds^i}{dt} = D\left(s_{in}^i - \frac{k_y}{k_y^i} s_{in}\right) \tag{24}$$

Equation (24) states that $ds^i/dt \approx 0$ only when $s_{in}^i = (k_y/k_y^i)\, s_{in}$. Therefore, the concentrations of other nutrients are not always constant.

5.4 Effect of Maintenance Energy

When $m \neq 0$ in a constantly-fed-batch culture [f_v = const. in Eq. (3)], the microorganisms do not grow linearly in the succeeding growth phase, while m only minorly affects either vx, μ, or x in the early exponential growth phase (Fig. 2). Since the entire substrate fed into a bioreactor is consumed by the microorganism in the succeeding growth phase, the assumption can be maintained that d(vs)/dt is negligible in Eq. (3):

$$\frac{1}{k_y}\frac{d(vx)}{dt} + mvx = f_v s_{in} \tag{25}$$

Integration of Eq. (25) under the initial condition: $vx = v_T x_T$ at $t = t_T$, where t_T is the transition time from the exponential to the succeeding growth phase, leads to:

$$vx = \frac{f_v s_{in}}{m} - \left(\frac{f_v s_{in}}{m} - v_T x_T\right) \exp\{-mk_y(t - t_T)\} \tag{26}$$

Equation (26) indicates that the total cell mass in the bioreactor vx approaches a constant value of $f_v s_{in}/m$ as $t \to \infty$ and that no growth occurs when $f_v s_{in} \leqq m v_T x_T$. Thus, the maximum total amount of biomass in the bioreactor $(vx)_{max}$ is given by:

$$(vx)_{max} = f_v s_{in}/m \tag{27}$$

These considerations will be useful in determining how much substrate should be added to prolong the stationary growth phase where secondary metabolite or non-growth associated metabolite formation predominates.

Equations (26) and (27) can be used to experimentally determine the value of m of the microorganism, as suggested by Marr et al. [96]

5.5 Exponentially-fed-batch Cultures

Under ideal conditions microorganisms will grow exponentially. If the feed rate of the growth-limiting substrate is increased in proportion to the exponential growth rate, it is possible to maintain a high rate of growth for a long time, while keeping the substrate concentration in the culture broth at a constant level. In the system represented by Eqs. (1) to (3) and (5) to (8), the necessary and sufficient conditions for exponential growth:

$$vx = v_0 x_0 \exp(\mu t) \tag{28}$$

can be given by the following equations [11,25,97,98], in which μ is constant and the volume is variable:

$$\left.\begin{aligned} s_0 &= g^{-1}(\mu) \\ f_v &= \frac{\mu/k_y + m}{s_{in} - s_0} v_0 x_0 \exp(\mu t) \end{aligned}\right\} \qquad (29)$$

For constant volumes:

$$\left.\begin{aligned} s_0 &= g^{-1}(\mu) \\ f_s &= (\mu/k_y + m)\, v_0 x_0 \exp(\mu t) \end{aligned}\right\} \qquad (30)$$

The required flow rate (volumetric or mass) must be increased exponentially with time, and therefore this mode of fed-batch culture is called exponentially-fed-batch culture (operation) [25].

The main features of the exponentially-fed-batch culture are: (a) the constant substrate concentration; (b) the voluntary control of the specific growth rate of microorganisms by external variation of the exponent μ of the feed rate, provided that $\mu \leqq \mu_{max}$. The exponentailly-fed-batch culture is, therefore, somewhat similar to the chemostat with constant dilution, although it is, in fact, a modification of the batch culture [25,97].

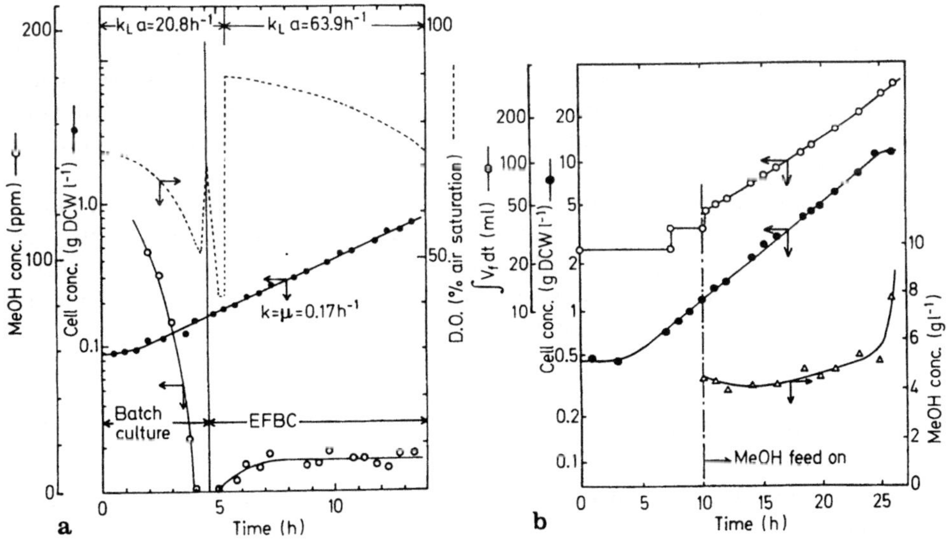

Fig. 10a and b. Time courses of exponentially-fed-batch cultures with an undiluted methanol supply in (**a**) stable and (**b**) unstable regions [11,25]. Conditions: (**a**), $\mu = 0.17\ h^{-1}$, $x_0 = 0.15\ g\ l^{-1}$, $k_y = 0.3\ g\ g^{-1}$, (**b**), $\mu = 0.16\ h^{-1}$, $x_0 = 1.16\ g\ l^{-1}$, $k_y = 0.3\ g\ g^{-1}$

Since the feeding of the substrate in the exponential fed-batch culture is not feedback controlled around a set-point but is increased according to a predetermined program, it is important from a practical viewpoint to know how stable the system will be if one of the state variables x, s, or v accidentally fluctuate slightly. From our theoretical analysis by the method of perturbation, it is concluded that the system is stable if $dg(s)/ds > 0$ but unstable if $dg(s)/ds \leqq 0$ [25]. We have varified this conclusion in an exponentially-fed-batch culture of methanol-assimilating bacteria. Figure 10(a) shows experimental results in the stable region, and Fig. 10(b) in the unstable region.

The exponential fed-batch culture is the most suitable means of obtaining maximum amounts of cell mass in the shortest time with a substrate that exhibits lag-time elongation and growth-inhibition at high concentrations. Generally, high yields of an intracellular metabolite can optimally be obtained in a short time by performing an exponential fed-batch culture at $\mu \approx \mu_{max}$ at a highest possible cell concentration.

5.6 Optimized Fed-batch Operations

The semi-batch operations in microbial processes with constant and exponential feed rates are most practical in terms of simplicity and as it corresponds very well to the nature of microbial growth. Yet, whenever a microbial process is operated semi-batchwise under real economic circumstances the feed rate of the substrate should be changed so that a judiciously chosen optimization criterion may be met. As in many other processes, there are several possible optimization criteria. These are called objective functions in process systems engineering. They include the maximization of yield (the total amount of the desired metabolite in a bioreactor at the harvest time), the productivity (the yield divided by the running period) or profit, and the minimization of production costs. Formerly, processes were optimized on a purely empirical basis. In the last 15 years, however, a number of applications of modern mathematical optimizations for microbial processes have been reported involving optimization of the feeding mode [99]. In this section, we review the application of rigorous and heuristic mathematical optimization methods to the fed-batch operation in microbial processes and discuss the important developments and trends in this field.

In 1974, Fishman et al. [54] reported an optimal strategy for feeding glucose in fed-batch penicillin production. They developed a kinetic model, which very interestingly took consideration of cell age of the culture and its effect on product synthesis. Growth was assumed to follow the Monod kinetics, which was modified for the presence of an inhibitory intermediate metabolite, as follows:

$$\frac{dx}{dt} = \frac{\mu_{max} sx}{k_s + s} - b_1 ix \tag{31}$$

Glucose assumably was consumed primarily by biomass in a growth-associate manner:

$$\frac{ds}{dt} = -\frac{1}{k_y}\frac{\mu_{max} sx}{k_s + s} + \frac{f_s}{v} \tag{32}$$

The inhibitory metabolite formed according to:

$$\frac{di}{dt} = b_2 \frac{\mu_{max} sx}{k_s + s} + b_3 ix \tag{33}$$

Penicillin production was modeled by:

$$\frac{dp}{dt} = q(\bar{\Lambda})\,x \tag{34}$$

where q is a function of the mean age $\bar{\Lambda}$ of the culture, which was defined by:

$$\bar{\Lambda}(t) = \frac{x_0\bar{\Lambda}_0 + \int_{t_0}^{t} x(t)\,dt}{x(t)} \tag{35}$$

$$q(\bar{\Lambda}) = a_0 + a_1\bar{\Lambda} + a_2\bar{\Lambda}^2 \tag{36}$$

The optimization problem is to find a program by which $f_s^*(t)$ changes within the interval $t_0 \leqq t \leqq t_f$ under the constraint $0 \leqq f_s(t) \leqq f_s^{max}$, so that the objective function, i.e., the final penicillin concentration, will reach its maximum at the optimal condition:

$$J = p(t_f) \rightarrow \max\,. \tag{37}$$

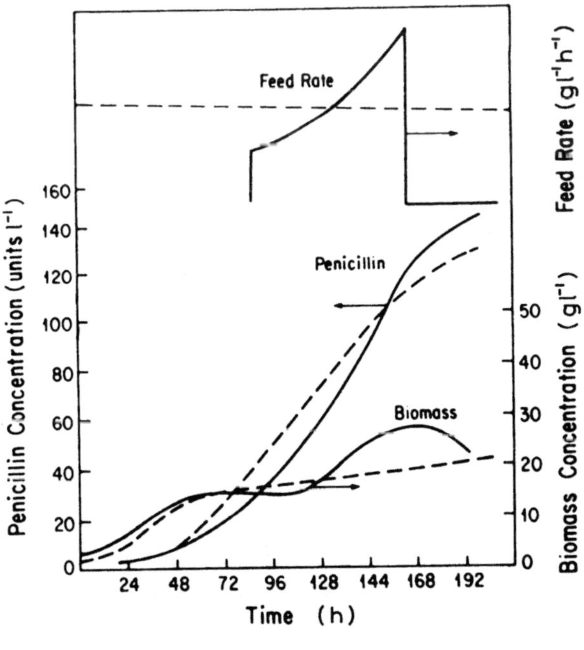

Fig. 11. Optimal control of glucose feed rate and changes in biochemical characteristics in fed-batch penicillin production [54]

The application of Pontryaghin's maximum principle to this problem yielded a bang-singular-bang profile for the rate of glucose feeding as shown in Fig. 11. The profile of the state variables under optimum control are shown in the same figure. An increase of about 7% in the final penicillin concentration was obtained by the optimal feeding profile instead of the constant feeding rate. The dilution effect through the glucose addition on the state variables, as noted by Constantinides [99], is negligible if a concentrated solution is fed. The solubility of glucose in water is as high as 500 g l^{-1} at room temperature.

Soon after the work of Fishman et al. [54], Ohno et al. [69] formulated a mathematical model for semi-batch processes and applied Green's theorem to solve the maximization problem. The system equation they formulated is identical to Eqs. (1), (2), (3), (5), and (8) with k_y = const. and m = 0. The objective function is the productivity defined by:

$$J/t_f = v(t_f)p(t_f)/t_f \tag{38}$$

at a free harvesting-time t_f. Their optimization problem was to find $f_v(t)$ maximizing Eq. (38) with the initial and final conditions: $v(0) = v_0$, $s(0) = 0$, $x(0) = x_0$, $v(t) = v_f$ and the inequality constraint: $0 \leqq f_v \leqq f_v^{max}$. Having established two theorems they proposed an iterative procedure to find the optimal trajectory that maximizes the productivity given by Eq. (38). They applied their optimization procedure to a lysine production process by an auxotrophic mutant with the following kinetic equations:

$$\mu = 0.125s\,, \qquad k_y = 0.135\,, \qquad q_p = -384\mu^2 + 134\mu \tag{39}$$

Their numerical results revealed that a constant feed rate, suboptimal but easily realizable, is comparable to the optimal one.

In the optimization method based on Pontryaghin's maximum principle the mathematical formulation is reduced to a rather difficult 'two-point, boundary value problem'. The optimization method based on Green's theorem is simpler but generally applicable only to the problem of two state-variables with both terminal states fixed. If a state variable is not fixed, an iterative procedure is needed. Since these two methods involve complicated calculations, we applied engineering principles in seeking a simpler method, even though it might entail sacrifice of some mathematical rigour [100]. The initial condition and the inequality constraint imposed on the mathematical model are:

$$x(0) = v_0\,, \qquad v(0) = v_0\,, \qquad vx \leqq (vx)_{max}\,, \qquad f_v \geqq 0 \tag{40}$$

Two objective functions, i.e., Eq. (41):

$$J = v_f p_f = \int_0^{t_f} h(\mu)\, vx\, dt \tag{41}$$

and Eq. (38) were used for our method. Inspection of the mathematical model reveals several characteristics: (a) x always appears as its product with v; (b) the objective function (41) depends only on μ and vx, and contains explicitly neither t nor f_v; (c) one of the system equations, Eq. (1), depends only on μ and vx, and contains explicitly neither t nor f_v; (d) v can be expressed as a function of (μ, vx):

$$v = \frac{\{g^{-1}(\mu_0) - s_{in}\}\, v_0 - \int_0^t (\mu/k_y + m)\, vx\, dt}{g^{-1}(\mu) - s_{in}} \tag{42}$$

Based on these characteristics the problem can be simplified, if the constraint $f_s \geqq 0$ is set aside, by regarding μ as an operating variable and vx as the state variable instead of solving the original problem directly. The problem now is to find μ(t) that maximizes Eqs. (38) or (41) using Eqs. (1) and (5) with the initial condition $vx = v_0x_0$ and the inequality constraint $vx \leqq (vx)_{max}$. This problem is easily solved analytically by the continuous maximum principle. The general result includes three possible solutions. These are illustrated by the trajectories on the μ/μ_{max} vs. vx/v_0x_0 phase plane in Fig. 12. It must be emphasized that our simple mathematical procedure to optimize the feed rate of the growth-limiting substrate for metabolite production by a fed-batch operation is applicable to any system having an arbitrary functional formula of h(μ).

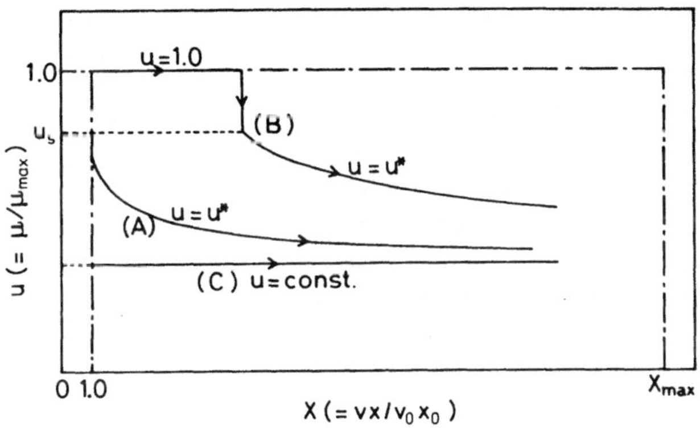

Fig. 12. Three possible optimal trajectories of μ/μ_{max} vs. vx/v_0x_0 in the simplified optimization procedure [100]

Application of rigorous optimization methods to the fed-batch microbial reaction, three of which have briefly been reviewed here, requires the development of mathematical models that describe adequately the behavior of the system under a variety of conditions, and give quantitative relationships between the state variable and the feed rate of the limiting substrate. This requirement, however, is not always fulfilled in industrial microbial processes, where cheap, natural but complex media are frequently used and feeding profiles are optimized by trial and error. This case is exemplified by a microbial production of β-amylase [75].

5.7 Predetermined Supply of Nutrients

In the early years, definite amounts of appropriate nutrients were almost exclusively supplied manually and intermittently to the bioreactor. In 1953, Dale et al. [101] reported 'the increment-fed shaker system', a cam-driven feeding assembly fitted to a laboratory shaker. The following year, a more complicated continuous feeder for Erlenmeyer flasks on a rotary shaker was reported for penicillin fermentation [45]. A rotary positive displacement pump fitted to a timer was introduced by the same researchers to vary feed rates with bench-scale fermentors [43,44]. This pump was designed and used to add glucose, lactose, or sucrose to a medium to improve penicillin production. Nowadays, this type of pump is called a peristaltic pump or tubing pump and is commonly used for adding nutrients aseptically with various modes of control. The simplest control is on/off control operated manually or by a timer. We were the first to design a rotating drum controller as a feed rate programmer connected with a tubing pump [25]. The profile of flow rate change with time can be controlled arbitrarily by changing the shape of the black paper affixed to the drum surface; the pump operates of 5–10 second on/off intervals. Therefore, the addition is microscopically intermittent but macroscopically continuous, and the time-averaged feed rate changes smoothly. The next apparatus we designed was an automatic curve tracer again connected to a peristaltic pump [75]. A photoprobe traces the black-and-white boundary of a sheet of paper. By changing the shape of the boundary, the tracing speed, the gear head of the pump, the diameter of the tubing, as well as the substrate concentration in the feed solution, we can obtain a pre-determined pattern of feeding and a very wide range of mass feed rate. The addition is perfectly continuous. Other techniques have been reported such as the exponential gradient generator, which allows only exponentially increased flow rates [102], the gradient-feed method, in which the feed rate is constant but the concentration of the nutrients in the feed increases linearly with time [26], and an automatically controlled feeding through a programmed automatic pump with timers and relays [27]. To overcome problems of high cost and low feed rates in shake flask experiments, the diffusion capsule [103] and coated tablets [49] have been suggested, both of which retard the rate of diffusion of the substrate into the liquid medium.

6 Fed-batch Operation with Automatic Feedback Control

In the fed-batch operations without feedback control, where substrates are added according to a pre-determined fixed schedule, it is difficult to cope with unforeseeable and unfavorable conditions of the microbial reaction during operation. Time courses may not always follow the expected profiles. It is desirable, therefore, to incorporate a negative feedback control into the feeding system if possible. The most primitive means is to have a technician on site performing the control manually, which is possible in some of the rather slow microbial processes. Automatic control, however, is favored over manual control as of higher precision and rationalization. The fed-batch (or semi-batch) operation with automatic feedback control can be divided into two categories: those with indirect feedback control and those with direct feedback control.

6.1 Fed-batch Operation with Indirect Feedback Control

This mode of operation utilizes an observable parameter that is closely connected with the course of the microbial process. Several such feedback parameters have been reported, including dissolved oxygen concentration (DO), respiratory quotient (RQ), hydrogen ion concentration (pH), metabolite concentration, and partial pressure of CO_2 in the exhausted gas (P_{CO_2}). The mathematical relationship among the feedback parameters, the concentration of the fed substrate in the culture liquid, the feed rate of the substrate, the substrate uptake rate of the microorganism, and others should be known in order to carry out this type of fed-batch operation skillfully. Unfortunately, little work has yet been carried out along these lines.

6.1.1 Dissolved Oxygen (DO)

Utilization of DO in controlling substrate-feeding has a long history of over 25 years. The sterilizable, reliable DO electrode that is essential for the operation of the fed-batch culture with DO control is now commercially available. Since DO is apt to become a rate-limiting factor in industrial aerobic microbial processes, it is rational to control the substrate-feeding by DO as the feedback parameter.

The system equation of this operation consists of Eqs. (1), (3), (8), and the differential balance of DO:

$$\frac{d(v[DO])}{dt} = k_L av([DO]^* - [DO]) - q_{O_2}(vx) \tag{43}$$

where q_{O_2} is the specific respiration rate or oxygen uptake rate of the microorganism. Hospodka [16] discussed the dynamic behavior of this feeding system to keep DO at the set level (DO-stat) on the assumption that $v = \text{const}$. In this case, Eq. (43) is reduced to:

$$\frac{d[DO]}{dt} = k_L a([DO]^* - [DO]) - q_{O_2}x \tag{44}$$

The system works only in the substrate-limited region and $[DO] > [DO]_{cri}$, where $[DO]_{cri}$ is the critical oxygen concentration. It may be possible to operate the system below $[DO]_{cri}$ under oxygen limitation, but the insufficient sensitivity and poor stability of the oxygen electrode do not allow this in practice. In the substrate-limited but oxygen-unlimited condition, q_{O_2} is independent of [DO]:

$$q_{O_2} = \frac{q_{O_2}^{max} s}{k_s^0 + s} \approx \frac{q_{O_2}^{max}}{k_s^0} s \tag{45}$$

From Eqs. (44) and (45), we receive the operating equation for the constant DO:

$$[DO] = -\frac{q_{O_2}^{max} x}{k_L a k_s^0} s + [DO]^* \tag{46}$$

In reference to Eq. (46), the control of substrate-feeding with DO-stat is described qualitatively as follows: If the value of DO increases slightly owing to a decrease in s caused by assimilation, the controller opens the valve and adds the substrate, and s increases with a resulting decrease in the DO value at constant x and k_La. If DO reaches the set value of the controller, the valve shuts, s begins to decrease again, and the entire process is repeated. Thus, DO oscillates around the set value, and the amplitude of the oscillation depends on the quantity of substrate added at a time. To decrease the oscillation, the system should be modified so that only a part of the substrate feed is controlled, the main portion flowing in continuously. Alternatively, proportional control of a pump may be used. The growth rate is related to q_{O_2}. If $q_{O_2} = \mu/k_y^0$ at the DO-stat, which is derived from the assumption that cell yield based on the consumed oxygen k_y^0 is constant, then the following equation is derived from Eq. (44):

$$\mu = k_y^0 k_L a([DO]^* - [DO])/x \qquad (47)$$

From Eq. (47), we can understand that the growth rate decreases gradually owing to the increase in x when k_La and [DO] are kept constant. It is therefore necessary to increase k_La or [DO]* to keep μ constant, if such is desired.

A number of papers have appeared concerning controlled substrate feeding mediated by DO-stat [15,16,17,52,105,106]. Kobayashi et al. [31,32,105] have developed two types of DO-stats in which agitation speed and aeration rates were used as control elements. These systems required some manual operations during cultivation. In

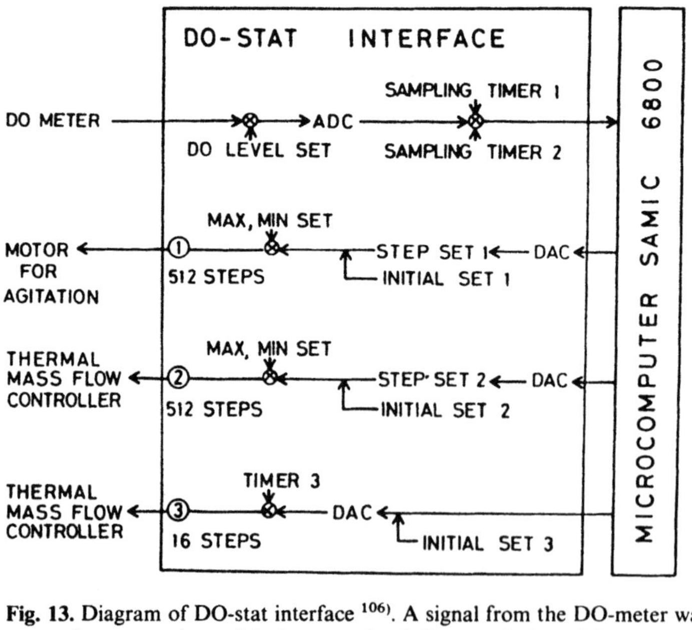

Fig. 13. Diagram of DO-stat interface [106]. A signal from the DO-meter was transmitted to the computer after comparison with the preset level. Outputs 1, 2, and 3 were connected with the motor for agitation and the termal mass flow controller for air and oxygen, and could be changed in 512, 512, and 16 steps, respectively

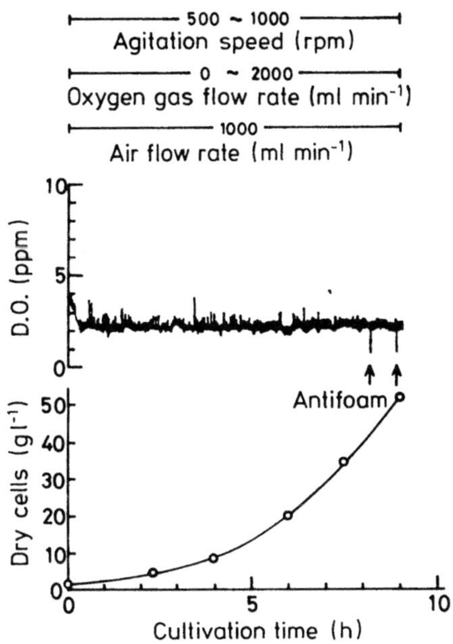

Fig. 14. Data from a cultivation of *Candida brassicae* in a fed-batch process with a computer-aided DO-stat [106]. When agitation speed reached the preset minimum or maximum level, the oxygen flow rate was changed automatically. The air flow rate was kept constant (1 vvm)

order to avoid manual operation of the DO-stat system, a microcomputer has been used and an apparatus interfaced with a microcomputer was introduced to maintain DO at a constant level [106]. Agitation speed and air/oxygen flow rates were automatically varied according to instructions stored in the microcomputer (Fig. 13). One of their results is shown in Fig. 14. The control system they have developed will have wide applications for various fed-batch operations.

6.1.2 Respiratory Quotient (RQ)

As early as 1959, Dietrich [107] suggested the use of RQ as a control parameter for yeast production. Since then, substrate feeding controlled with RQ as a feedback parameter has been used exclusively for baker's yeast propagation. This is because the four types of yeast metabolism (oxidation of glucose, aerobic fermentation resulting in the formation of ethanol, oxidation of both glucose and ethanol, and assimilation of ethanol by oxidation) are related to the level of the RQ-value. The aim is to minimize the formation of ethanol (by-product) and achieve the highest possible cell yield from sugar by avoiding both over- and under-feeding of the sugar. The RQ value is calculated by:

$$RQ = r_{CO_2}/r_{O_2} \tag{48}$$

where r_{CO_2} and r_{O_2} are the volumetric carbon dioxide evolution rate (m mol l^{-1} h^{-1}) and volumetric oxygen uptake rate (m mol l^{-1} h^{-1}), respectively. r_{CO_2} and r_{O_2} in an actual microbial reaction are computed from carbon dioxide and oxygen concentrations in the exit and inlet gases and the air flow rate.

Aiba et al. [18] reported yeast production by manually changing the sugar feed rate stepwise so as to maintain the RQ value within the range of 1.0 to 1.2 throughout the cultivation. A linear relationship between RQ and r_{EtOH} was derived by Wang et al. [19]:

$$r_{EtOH} = \{(RQ) - (RQ)_0\} r_{O_2} \tag{49}$$

where r_{EtOH} is the volumetric ethanol production rate (m mol l^{-1} h^{-1}) and $(RQ)_0$ is the RQ value in the absence of ethanol formation. From Eqs. (48) and (49) we derive:

$$r_{EtOH} = r_{CO_2} - (RQ)_0 r_{O_2} \tag{50}$$

They noticed that the control of yeast production was not satisfactory in the experiments where only the RQ value was used to control molasses addition and they developed a combined anticipatory/feedback control strategy as shown in the following equations [13]:

$$f_s = \frac{\mu v x}{k_y s_f} \tag{51}$$

$$f_{ss} = f_s[1 - k_c\{r_{CO_2} - (RQ)_0 r_{O_2}\}] \tag{52}$$

Equation (52) expresses a negative feedback of f_s. In these equations, f_s is an anticipated volumetric feed rate of molasses computed by Eq. (51) from the indirect cell measurement technique together with specified values of the desired specific growth and cell yield. f_{ss} is the corrected molasses flow rate and k_c is the controller gain. $(RQ)_0$ was set between 1.05 and 0.95. One of their computer-controlled experiments

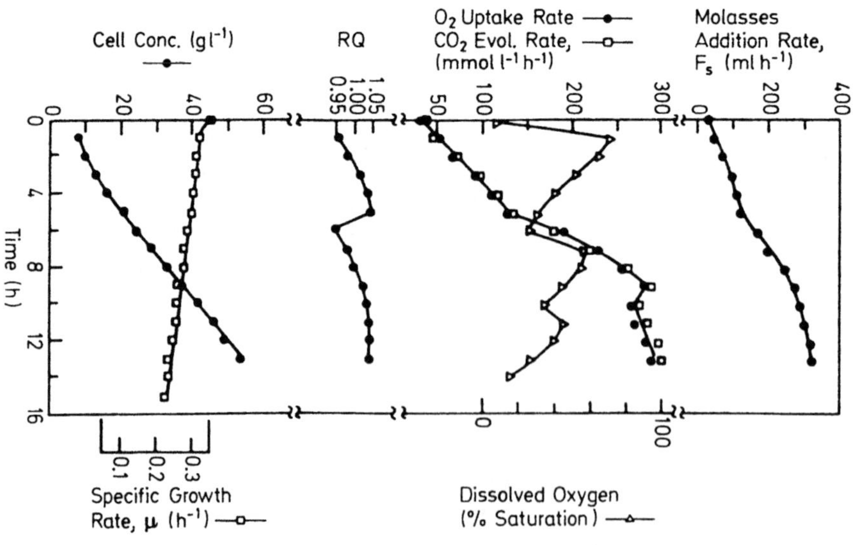

Fig. 15. Performance of a fed-batch yeast production process with RQ control by a computer [13]

is shown in Fig. 15. The ultimate test of the control system showed its effectiveness in overcoming problems of oxygen starvation, molasses quality, and variable inoculum size.

6.1.3 Hydrogen Ion (pH)

The fed-batch mode of operation with the pH as the feedback parameter is based on the phenomenon that when a medium component is consumed during cultivation, pH values deviate (often decrease) from a set value. Alternatively, it is useful for the controlled feeding of substrate, such as acetate which cause changes of the pH within the culture [1]. Nishio et al. [30] reported a semi-batch culture where a methanol-ammonia mixture was fed in response to a direct signal of pH change. Gottvaldovà et al. [78] reported an enhancement of cellulase production by *Trichoderma viride* using a carbon/nitrogen double fed-batch culture with pH-stat.

Successfull application of this technique should make it possible to establish a quantitative relationship between the rates of substrate feeding and the pH change, as well as of cell growth. Synthetic or semi-synthetic medium is preferable for analyzing this relationship. Even so, the response of pH to the addition of substrate is usually not very fast.

6.1.4 Partial Pressure of Carbon Dioxide in the Exit Gas (P_{CO_2})

Assimilation of a carbon source by microorganisms is always accompanied by CO_2 evolution, and feeding of a carbon source, negatively feedback controlled by the rate of CO_2 evolution, will result in a limited concentration of the carbon source in the medium. The CO_2 evolution rate is computed by the CO_2 concentration in the outlet gas that is measured continuously with an infra-red CO_2 analyzer. The pH of the culture should be controlled accurately. A poor pH control system leading to fluctuations of the pH might lead to short-term changes in CO_2 evolution, since CO_2 solubility is very pH dependent. This feedback control strategy was adopted by both Suga et al. [76] and by Allen et al. [77] for the production of *Trichoderma reesei* cellulase. The control system of Suga et al. had two units, an on/off control for feeding the substrate to the bioreactor and a programmer that would increase the position of the on/off value exponentially. In the computer-aided control system of Allen et al., substrate (cellulose) addition was proportional to the negative deviation from the present CO_2 value (proportional control); no control was applied for positive deviation from the preset value. Mou et al. [56] succeeded in controlling μ by monitoring P_{CO_2} and applying computer-aided on-line mass balancing in fed-batch penicillin production.

6.1.5 Concentration of Metabolites

If an extracellular metabolite is formed from the fed substrate and its concentration can be measured continuously by a sensor, we can control the substrate-feeding in such a way that it does not accumulate. Alternatively, if it is possible to continuously measure the concentration of a metabolite whose formation rate is closely related to the concentration of the fed-substrate, this might also be used as a control parameter. A good example of the former case is ethanol in baker's yeast production. The glucose feed rate is controlled to maintain a constant ethanol concentration.

The ethanol concentration in the culture liquid can be measured directly by the porous Teflon tubing method [108], or indirectly from its vapor concentration in the exit gas [22,23]. Dairaku et al. [20] first studied the maintenance of the ethanol concentration at a level of 150 ppm through a system which combined the porous Teflon tubing method with a flame ionization detector (FID) and an on/off or PID controller. The addmissible upper limit of the ethanol concentration to avoid ethanol formation was about 2000 ppm. One of difficulties they faced in regulating the fed-batch culture skillfully by means of normal PID feedback control arose from the fact that cell concentration increases with time. The system has never reached a steady state. Control parameters had to be readjusted in accordance with the change of the system dynamics. To overcome this difficulty they developed an advanced control system that utilizes a simple mathematical model of the increase in cell mass and PID feedback control incorporating automatic parameter tuning by a microcomputer [21]. In the advanced control model, the actuating signal Z is given by:

$$Z = Z_{est} + \frac{1}{K_p}\left(\varepsilon + \frac{1}{T_i}\int_0^t \varepsilon \, dt + T_d \frac{d\varepsilon}{dt}\right) \tag{53}$$

where Z_{est} is the estimated actuating signal and the latter part is the PID feedback control element. The control parameters K_p, T_i, and T_d are called the proportional band, integral time constant, and differential time constant, respectively. Z_{est} is obtained from the linear relationship between the glucose feed rate f_v and the actuating signal Z: $f_v = aZ_{est} + b$, where $0 \leqq Z_{est} \leqq 1$, together with Eqs. (3) and (6) ($m = 0$) under the assumption that $d(vs)/dt \approx 0$:

$$Z_{est} = \frac{1}{a}\left(\frac{\mu}{k_y s_f} vx - b\right) \tag{54}$$

The effects of various levels of x and s_f on the values of individual control parameters were analyzed experimentally. K_p was found to be proportional to both x and s_f:

$$K_p = K_{ps} \frac{x}{x_s} \frac{s_{in}}{s_{ins}} \tag{55}$$

T_d was almost constant irrespective of changes of x and s_f. T_i was found to be four times that of T_d. vx is estimated from Eq. (3) ($m = 0$) under the assumption: $d(vs)/dt \approx 0$.

$$vx = v_0 x_0 + k_y \int_0^t f_v s_{in} \, dt \tag{56}$$

From Eqs. (54), (55), and (56) the advanced control algorithm becomes:

$$Z = \frac{1}{a}\left\{\frac{\mu}{k_y s_{in}}\left(v_0 x_0 + k_y \int_0^t f_v s_{in} \, dt\right) - b\right\} + \frac{1}{k_{ps}}\left(\frac{x_s}{x}\right)\left(\frac{s_{ins}}{s_{in}}\right)\left(\varepsilon + \frac{1}{T_i}\int_0^t \varepsilon \, dt + T_d \frac{d\varepsilon}{dt}\right) \tag{57}$$

The control system was tested experimentally by adding glucose pulse-wise or changing s_f stepwise. One of the results of the baker's yeast fed-batch culture with the proposed control system is shown in Fig. 16. Control of disturbances was very well performed by the system. Precise estimation of x_0 is necessary for good performance with this control system. Better results will be obtained by taking the volume change into account.

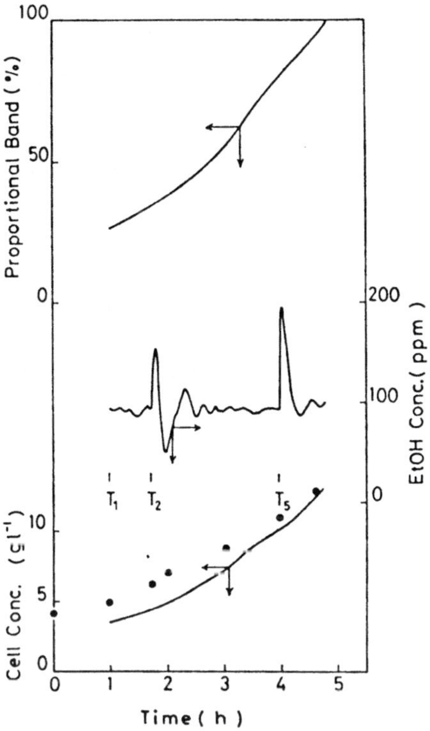

Fig. 16. Results of the advanced control system with large disturbances [21]. Glucose was added pulse-wise at T_2 and T_5. The estimated cell concentration is expressed as a solid line, and the measure cell concentration is expressed as (●)

6.2 Fed-batch Operation with Direct Feedback Control

In the semi-batch operation with a direct feedback control, which is usually automatic and negative, the concentration of the fed-substrate in the culture liquid measured continuously (or intermittently if continuous measurement is not feasible) is utilized directly as a feedback parameter (control index). It is not always necessary to fix a value of its concentration; this may be changed in the optimized pattern during operation. A semi-batch culture with a direct feedback control is superior to all feeding modes mentioned above for circumventing any unforeseen deterioration of the culture during operation. Furthermore, this mode is also the most desirable considering that the concentration of the substrate affects the physiological state of the microorganism. As we have seen earlier, semi-batch culture without feedback control and with indirect feedback control have been extensively studied. However, very few studies have been done on fed-batch culture with direct feedback control.

probably owing to the difficulty of continuously measuring the concentration of biochemical substances in situ.

If the fed substrate is volatile and its partial pressure in the exhaust gas is in equilibrium with its concentration in the culture liquid and the partial pressure (gaseous concentration) is continuously measurable, then we can use the partial pressure as a control index. Reuss [29] has carried out research along these lines. Yano et al. [109] reported a semi-batch culture by feeding methanol whose concentration was measured intermittently by means of a silicone rubber tubing sensor combined with a gas chromatograph. The methanol concentration in the culture liquid was controlled manually according to information on the observed peak heights of the chromatogram. Soon thereafter Dairaku et al. [108] demonstrated the superiority of porous Teflon tubing for continuous concentration measurements. Yamanè et al. [110] found its mass-transfer and dynamic characteristics useful for designing the sensor. Then, they constructed an automatic feed control incorporating a porous Teflon tubing sensor, and carried out semi-batch culturing of a strain of yeast in a mineral salt medium by feeding pure ethanol as the sole carbon source while maintaining its concentration in the culture liquid automatically at a constantly set level by means of the control apparatus [35]. The control algorithm was a proportional plus differential (PD) control system. The parameters of these control elements were determined by material balance. An actual profile of the ethanol concentration during the cultivation

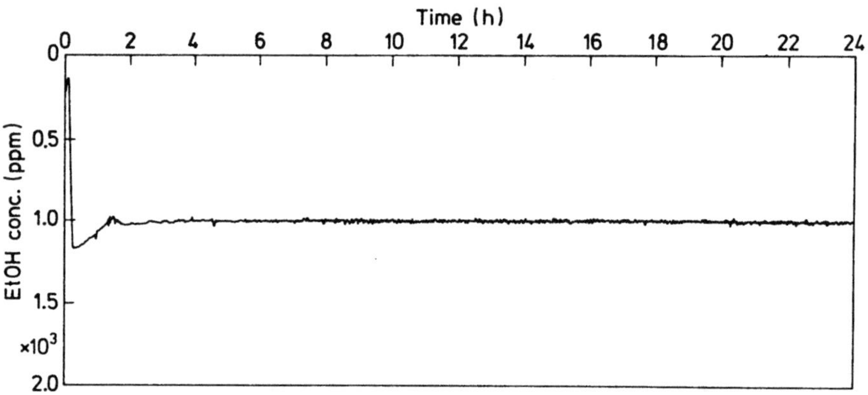

Fig. 17. Time course of actual ethanol concentration in a fed-batch culture with direct feedback control ($s_s = 10^3$ ppm) [35]

of *Pichia farinosa* is shown in Fig. 17. The control system was satisfactory when the set levels were greater than 10^3 ppm, but an offset arose at the 10^2 ppm level. Better control should be achieved with the advanced control system proposed by Dairaku et al. For semi-batch cultures with direct feedback control, with s_{in} = const., the actuating signal Z is given by:

$$Z = \frac{1}{a}\left(\frac{\mu}{k_y s_{in}} vx - b\right) + \frac{1}{K_{ps}} \frac{(vx)_s}{vx}\left(\varepsilon + \frac{1}{T_i}\int_0^T \varepsilon\, dt + T_d \frac{d\varepsilon}{dt}\right) \tag{58}$$

where vx is estimated by Eq. (56).

As a more sophisticated semi-batch operation with direct feedback control, in which the concentration of the fed-substrate in the culture liquid is varied automatically and optimally during the operation, Kishimoto et al.[67] have proposed on-line optimal control of fed-batch processes in glutamic acid production from ethanol by *Brevibacterium devaricatum* (Fig. 18). The algorithm they developed is a combina-

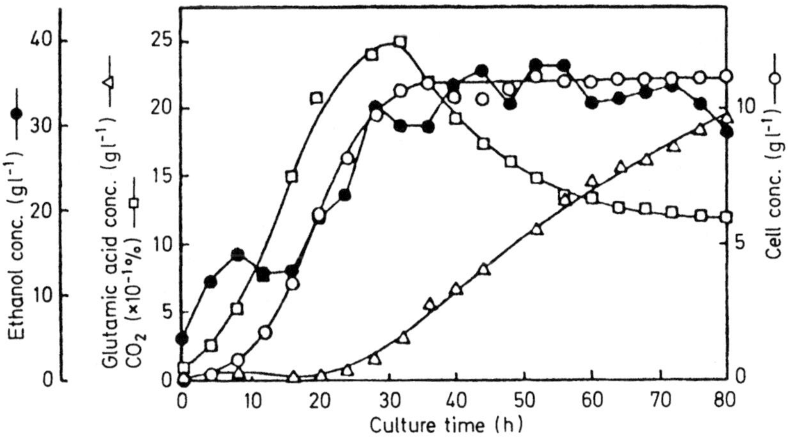

Fig. 18. On-line adaptive optimal control of a fed-batch process for glutamic acid production from ethanol by *Brevibacterium devaricatum*[67]

tion of optimization by dynamic programming and linear prediction of parameters by multiple regression analysis[111]. The experimental data used for multiple regression analysis were obtained from time courses of several fed-batch cultures, in each of which the ethanol concentration was either kept constant or increased suddenly during cultivation. The optimization procedure does not involve deterministic kinetic equations and has a high degree of flexibility to cope with the poor reliability and reproducibility of experimental data that are often encountered in microbial processes.

7 Extension and Application of Fed-batch Operations

7.1 Cultivation of High Concentrations of Microorganisms

Cultivation of microorganism of high cell concentrations, e.g., over 100 g of dry cells per liter, clearly increases volumetric cell productivity ($g\,l^{-1}\,h^{-1}$) and is also quite effective in decreasing the volume of waste water from the culture. The volumetric productivity of metabolites should also increase with the increase in cell mass, since the volumetric formation rate of metabolites is usually proportional to the cell concentration. As mentioned in Sect. 2.2, to obtain a high cell concentration in a batch culture while maintaining the optimal culture conditions, the carbon source must be fed gradually rather than added all at once prior to cultivation, irrespective of whether it is inhibitory or non-inhibitory to cell growth at its normal concentration.

Aerobic fed-batch culture of microbes to high cell concentrations with indirect feedback control of the substrate addition by DO-stat has been extensively studied by Kobayashi et al.[83, 105, 106] To increase the oxygen absorption rate, OAR, defined by:

$$OAR = k_La([DO]^* - [DO]) \tag{59}$$

The agitation speed of impeller and flow rates of air and pure oxygen gas were automatically changed. Near the end of the culture, the agitator speed was maximum and pure oxygen gas was supplied at a high flow rate. Details of their DO control system employing a microcomputer are reported in Ref.[10] With the DO-stat plus the microcomputer-aided control system, they obtained highly concentrated biomass in the fed-batch culture of *Protaminobacter ruber* (85 g l^{-1})[31], *Escherichia coli* (125 g l^{-1}), and *Candida brassicae* (138 g l^{-1})[33]. By fed-batch culture with DO-stat it was also possible to obtain highly concentrated sorbose (628 g l^{-1}) from sorbitol in a culture of *Gluconobacter suboxydans*[83] (Fig. 19), and highly concentrated biomass (153 g l^{-1}) of high lipid content (corresponding to 83 g l^{-1}) in an ethanol fed batch culture of *Lipomyces starkeyi*[84]. The common profile of the obtained cell growth

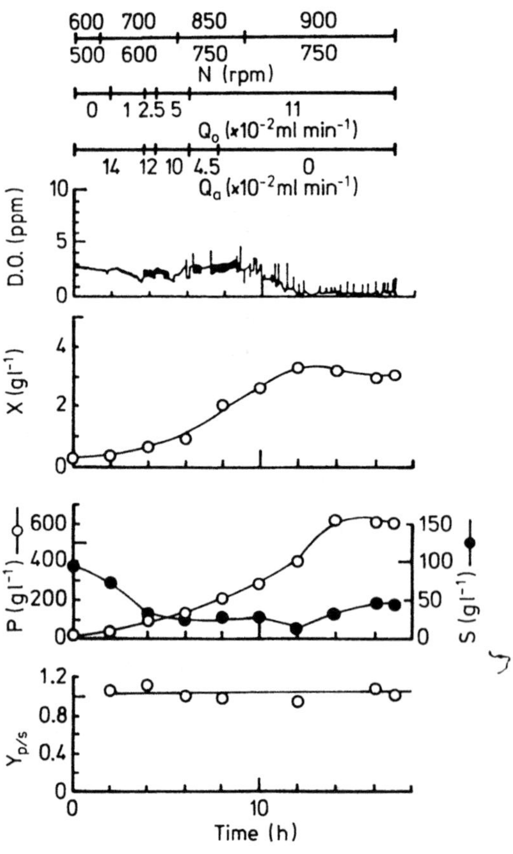

Fig. 19. Production of sorbose by *Gluconobacter suboxydans* in fed-batch culture with a computer-aided DO-stat[83]

curves shows that cell growth first was exponential when DO was unlimited (maintained at 2–3 ppm) and later decreased when DO dropped to zero. During the succeeding culture period, the cell growth was obviously limited by OAR. Therefore, the overall biomass productivity in fed-batch culture with DO-stat is generally limited by the oxygen transfer performance of the bioreactor.

In cultivating microorganism to a high concentration, all nutrients necessary for cell growth must be supplemented. Oxygen supply is automated by the DO-stat technique developed by Kobayashi et al. [83,105,106] Feeding of volatile carbon sources such as methanol or ethanol can be automatically feedback controlled with use of porous Teflon tubing sensors [35,108,110]. Ammonia added as a nitrogen source supplement can be automatically feedback controlled by use of a pH electrode-monitor (pH IC). Minerals can be automatically supplemented by adding them along with the main carbon source when the mineral composition of the microbe is unchanged [112]. If the ratio of the feed rate of the mineral solution is fixed to that of the carbon source at K_{MC} (—), the concentration of a mineral element C_{Mi}^{in} in the mineral reservoir is calculated by:

$$C_{Mi}^{in} = \frac{ky(10^3 \alpha_{dm}\gamma_{Mi} - \alpha_{H_2O}C_{Mi}^{sup})}{K_{MC}(10^3 \alpha_{dm} - k_y\alpha_{H_2O}S)} S_{in} + \frac{10^3 \alpha_{dm}(C_{Mi}^{sup} - k_y\gamma_{Mi}S)}{K_{MC}(10^3 \alpha_{dm} - k_y\alpha_{H_2O}S)} (K_{MC} + k_e f_s + k_N f_N - f_{vap}) \tag{60}$$

where the nomenclature is that of Ref. [112] Equation (60) is derived from the balance of the carbon source, the mineral element, and water, in which both the volume occupied by cells in the culture broth and the increase in the culture broth are taken into account. When the microorganism excretes metabolites inhibitory to cell growth, the combined technique of fed-batch culture and dialysis culture is effective for attaining high cell concentrations [113].

It should be of interest to note the upper limit of cell concentrations attainable by ideal fed-batch cultures. The physical upper limit may be the closest packing of cells. For *Protaminobacter ruber*, a rod-shaped microorganism, it was calculated as 410 g of dry cells l^{-1}. For the yeast *Candida brassicae*, which is almost spherical, the concentration of the cells packed by centrifugation (18,000 × g) is 258 g of dry cells l^{-1} [112]. This value is slightly higher than the cubic closest packing (the volume fraction is 74.2% in hexagonal closest packing) of uniformly sized spherical cells with 70% water, which is 223 [= 74.2/100 × 1,000 × (1 − 0.7)] g of dry cells per l. Such high cell concentrations are impossible to obtain due to the increase in viscosity of the culture broth. The viscosity of the culture broth increases sharply above 200 g of dry cells per l [31]. In view of fluid dynamics, the viscosity of the culture broth seems to determine the upper limit of the cell concentration.

7.2 Repeated Fed-batch Operation

When a portion of a "fed-batch culture" is withdrawn at intervals and the residual part of the culture is used as an inoculum for the next fed-batch culture, the whole system of operation is called "repeated fed-batch culture" [86]. Repeated filling and partial emptying to specified volumes result in a cyclic variable volume operation, from which this mode has alternatively been named "cyclic fed-batch operation" [114]. Informal reports suggest that in industry some antibiotics and baker's yeast are now produced by various types of repeated fed-batch processes developed on purely empirical bases. Pirt [86] developed an expression for the product concentration at the end of the n-th cycle as a function of a fixed fraction of residual volume at the end of each cycle when f_v = const., assuming that at all times the culture is in a quasi-steady state. By computer calculations Dunn et al. [90,114] simulated the biomass production rate of repeated fed-batch cultures with a constant f_v and they calculated the biomass productivity r_c for each reproducible cycle by:

$$r_c = x_f(1 - v_R/v_f)\, t_k \tag{61}$$

In order to obtain a 99% conversion of the added substrate before emptying the bioreactor, a constant-volume batch phase was added to each cycle. r_c vs. f_v curves were presented in nondimensional form for both Monod and substrate-inhibition kinetics, which showed that r_c follows the chemostat behavior but exhibits an earlier wash-out point and that the wash out depends on v_R/v_f. Later, Shioya et al. [115] proposed a graphical method to calculate cumulative biomass productivity, r_c', defined by:

$$r_c' = (x_f - x_0)/v_f x_k \tag{62}$$

for a repeated fed-batch culture with constant feed rate [89,114]. The biomass productivity of a repeated fed-batch culture was found to be higher than that of a repeated batch culture but lower than that of a chemostat. They also developed expressions for metabolite productivity of a repeated fed-batch operation at constant f_v for growth- and non-growth-associated metabolite formations (Luedeking-Piret kinetics). Weigand [116] considered optimal substrate feeding profiles and a culture removal policy which maximized cell productivity defined by the following equation for a repeated fed-batch culture when the cell yield was constant:

$$r_c'' = x_f(v_f - v_R)/t_k \tag{63}$$

Optimal operation was determined by the maximum principle. The general solution for μ being maximum (substrate-inhibition kinetics) consists of rapidly filling up to this maximum, followed by a singular input with an exponential policy that maintains the culture at the maximum μ, followed by a batch portion of the cycle where final conversion is achieved. If μ is monotonically increasing, then the optimal policy is repeated batch. Comparison of optimal, repeated fed-batch with other modes of operation in a single stirred tank for substrate-inhibited growth showed that cell productivity of repeated fed-batch exceeds both the repeated batch and the continuous cultures when the residual substrate concentration is low. All these papers [86,90,114–116] describe only theoretical studies without experimental verification. Furthermore, rapid instantaneous filling and draw off are assumed.

Mori et al. [117] proposed an expression for the biomass productivity of an n-cycle repeated fed-batch culture which takes into consideration the process time as well as culture time. The process is shown graphically in Fig. 20, and is given by:

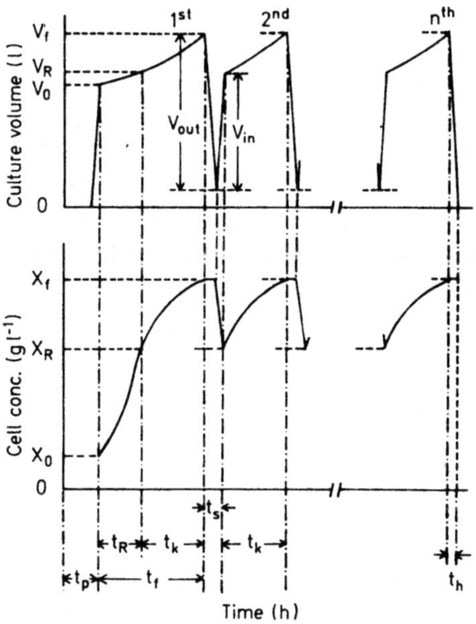

Fig. 20. Schematic diagram of a repeated fed-batch culture [119]

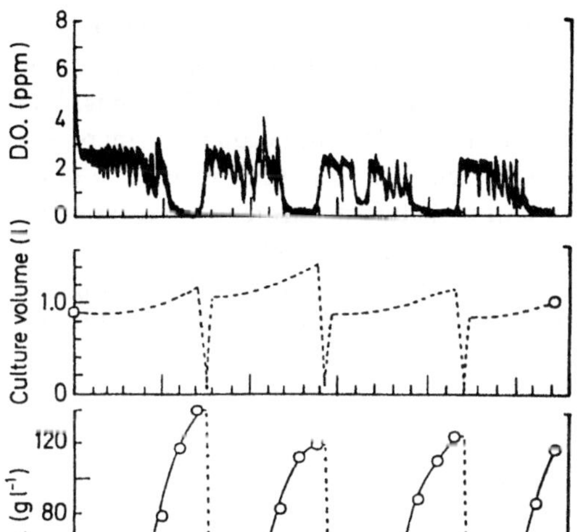

Fig. 21. Repeated fed-batch culture of *Candida brassicae* with ethanol feeding [119]. $v_{out}/v_f = 0.98$, $x_f = 140$ g l^{-1}, $v_f = 1.18$ l. Ethanol concentration was kept at 3.5 g l^{-1}

$$r_c''' = \frac{(n-1)\,x_f v_{out} + x_f v_f}{v_f\{t_p + t_1 + (n-1)(t_k + t_s) + t_h\}} \tag{64}$$

where t_p is the time necessary for preparation of the culture, i.e., cleaning and sterilizing the vessel, and filling it with sterilized medium for the first cycle. t_s and t_h are the times necessary to draw off the broth at the end of each cycle and after the final cycle, respectively. When the culture starts from x_R, then $t_1 = t_k$. They carried out several 4-cycle repeated fed-batch culture of *Candida brassicae* up to very high cell concentrations ($x_f \approx 160$ g of dry cells per l), supplying ethanol as the sole carbon-energy source. The ethanol supply was automatically feedback controlled to keep its concentration in the culture liquid constant (ca. 3.5 g l^{-1}) by use of a porous Teflon tubing sensor combined with a FID. One of their results is shown in Fig. 21. Cell productivities of fed-batch, repeated fed-batch, and chemostat cultures were compared on the basis of the ethanol feed rate. They increased in the order of fed-batch, chemostat, and repeated fed-batch cultures. The reason for the higher productivity of the repeated fed-batch culture than the chemostat appears to be a smaller OAR in the latter.

In addition to increased productivity, repeated fed-batch operation has the advantage that inocula need not be prepared for the second and subsequent cycles.

7.3 Start-up of Microbial Processes

The term "start-up" of a stirred-tank reactor identifies a sequence of operations that leads the reactor from an initial state to a specified final state. If the final state is a steady state of a continuous culture (chemostat), it is a start-up of a continuous culture. The optimal start-up of the chemostat leads to the steady state immediately after continuous operation is started. Yamanè et al. [118] established a time-optimal strategy for start-up of a chemostat by applying Miele's extremization method based on Green's theorem. The solutions are a batch culture for monotonically increasing growth kinetics and a exponentially-fed-batch culture followed by batch culture for substrate-inhibition kinetics. A time-suboptimal start-up strategy, an exponential fed-batch culture operated at the same μ as that of the chemostat, was proposed as more feasible than and physiological preferable to the time-optimal strategies (Fig. 22). The proposed time-suboptimal start-up strategy was applied to chemostat and

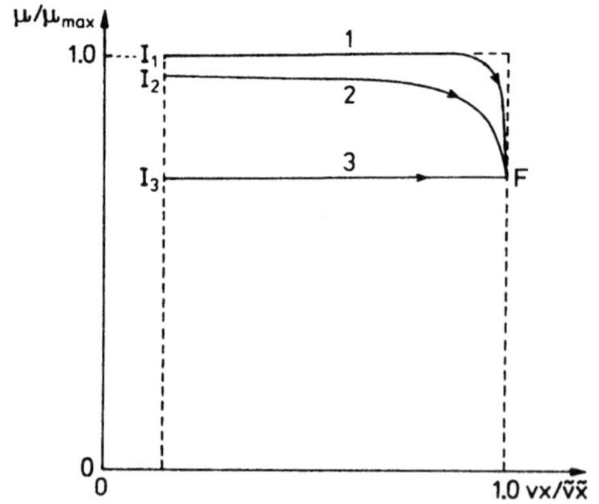

Fig. 22. Start-up of chemostat. ($vx/\tilde{v}\tilde{x}$, μ/μ_{max}) phase plane plots for (1) the exponentially-fed-batch culture followed by batch culture, (2) batch culture, and (3) the proposed time-suboptimal but physiologically optimal start-up (exponentially-fed-batch culture at $\mu = \tilde{\mu}$) [118]

nutristat of *Candida brassicae* that were carried out by feeding ethanol and mineral salt solutions [117,119].

Careful attention should be paid to the start-up of fed-batch operations. Excess substrate at the start of feeding prevents control of its concentration, while too small a quantity at the beginning of culture will prolong the lag-time. It is preferable to start the cultivation with a moderate substate concentration and to initially perform batch operation, and to begin feeding the substrate when the microorganism is growing actively but only when s is markedly decreasing. Figures 7 and 8 as well as 12 and 13 show the start-up of constantly- and exponentially-fed-batch cultures, respectively. In these cultures, substrate feeding can successfully be started just when the rapid rise in DO is observed. Dairaku et al. [120] examined the dynamics of an exponentially-fed-batch culture of baker's yeast by altering the glucose feed rate f_v so that μ changes stepwise in Eqs. (29) or (30). The observed response of μ was expressed by an equation of first-order plus time-lag, and with this model they established a time-optimal start-up policy of glucose feeding in good agreement with experimental results.

7.4 Applications to Microbial Physiology

Much work on microbial physiology has been done with chemostat cultures, which enables microorganisms to be grown in a strictly controlled environment. Fed-batch culture also fulfills some of the roles of chemostat culture in that it permits substrate-limited growth. The technique seems particularly suitable for the control of microbial processes at low growth rates. Fed-batch culture is technically easier than chemostat culture because such difficulties as wall growth, contamination by other strains, and maintenance of constant volume are involved in a long-term chemostat. For microbiologists, the elimination of these problems may make fed-batch culture more convenient than chemostat for studying the effects of such environmental factors as temperature and different nutrient limitations [86]. As early as 1963, Marr et al. [96] used, constantly-fed-batch cultures to study the maintenance energy requirement of *Escherichia coli*. More recently, Esener et al. [121] have underlined the power of fed-batch cultivation as an experimental tool for the simultaneous study of microbial kinetics and energetics with their extensive studies on fed-batch culture of *Klebsiella pneumoniae*. They also experimented with a unique feeding mode, decreasing the rate of feed linearly with time.

8 Concluding Remarks

It was shown that fed-batch operations can be classified with respect to the feeding mode according to Table 1, and each mode was mathematically described in detail. The characteristics of these feeding modes elucidated in the preceeding chapters should be helpful in deciding on a proper mode of fed-batch operation. It is advisable to test fed-batch operation after a regular batch operation and before engaging in continuous operations, since an increased yield or productivity can be expected with the fed-batch process, which in addition is much easier to operate than continuous operations.

In fed-batch operations with feedback control, setting up a reliable system equation to describe the microbial process in question is vital to the success of the operation. The development and improvement of accurate mathematical models, whether deterministic or stochastic, should be forwarded in future to allow operation of fed-batch cultures with optimized feed rate profiles.

In fed-batch operations with indirect feedback control, future trend will be to carry out the process with an optimized feed rate of the substrate(s) calculated by an on-line computer programmed with all obtainable information. To perform the fed-batch operation with direct feedback control, it is essential to measure the concentration of the fed substrate in the culture liquid. Unfortunately, sensors for direct monitoring of microbial processes on a commercial scale are presently very limited.

Much effort is needed to develop reliable, sensitive, and preferably, sterilizable sensors suitable for use in industrial microbial processes. Continuous measurement with rapid response would certainly be optimal; but even when measurement is intermittent, so-called sampled-data control will be possible if its time span is short enough.

The most preferable feeding mode will be "optimal feedback control", for which a reliable system equation should become firmly established. An optimal feeding strategy is mathematically determined by the rigorous optimization method, and the strategy is performed with an automatic feedback control by detecting directly the concentration of the fed substrate. Unfortunately, the current state of art in biotechnology and bioengineering seldom allows such fed-batch operations.

It is important, however, to realize that an increase in yield or productivity brought about by the application of a particular feeding strategy depends on the microorganism, the medium, the bioreactor, and the auxiliary equipment available for the particular process. Sometimes a carefully operated fed-batch culture with a constant feed rate of a substrate will suit the purpose of the operation; it may sometimes even be impossible to improve the results of the simplest feeding mode by introducing complicated and costly facilities including high-performance computers. Choice of the feeding mode is thus purpose-orientated. In fact, Bajpai et al. [122] suggested, from a careful investigation of four different strategies of sugar feeding, that the predicted, optimal productivities are more or less independent of the feeding scheme for the fed-batch process in penicillin production. In conclusion, the industrial chemicobio-engineer has to compromise in choosing a suitable feeding mode and balance the expected increase in yield or productivity with the economic situation.

9 Nomenclature

a_1, a_2, a_3	constants
b_1, b_2, b_3	constants
c	increment in culture volume per unit volumetric feed of substrate
C	dimensionless c
D	dilution rate, h^{-1}
DO, [DO]	dissolved oxygen concentration, mol l^{-1} or ppm

f	feed rate, $l\,h^{-1}$ or $g\,h^{-1}$
F	dimensionless f
i	concentration of inhibitory metabolite
J	objective function
k_y, k_y^i, k_y^o	cell yields based on growth-limiting substrate, i-species, and oxygen, respectively
k_s, k_s^o	saturation constants of growth and of oxygen consumption, respectivly, $g\,l^{-1}$
k_La	volumetric oxygen transfer coefficient, h^{-1}
K_s	dimensionless k_s
K_p	proportional band
m, m^i	maintenance coefficients, $g\,g^{-1}\,h^{-1}$
M	dimensionless m
OAR	oxygen absorption rate, $mol\,l^{-1}\,h^{-1}$
p	metabolite concentration, $mol\,l^{-1}$
q_{O_2}	specific oxygen consumption rate, h^{-1}
q_p	specific production rate of metabolite, h^{-1}
r	volumetric rate of formation and consumption, $mol\,l^{-1}$
r_c, r_c', r_c'', r_c'''	cell mass productivity, $g\,l^{-1}\,h^{-1}$
RQ	respiratory quotient
s, s^i	substrate concentration, $g\,l^{-1}$
S	dimensionless s
t	time, h
T	time constant
x	cell concentration, $g\,l^{-1}$
X	dimensionless x
v	volume of culture broth, l
V	dimensionless v
z	actuating signal

Greek letters

ε	relative deviation
Θ	dimensionless time
θ	time, h
$\bar{\Lambda}$	mean age of culture
μ	specific growth rate, h^{-1}
ϱ	density, $g\,cm^{-3}$

Subscripts

CO_2	carbon dioxide
d	differential
EtOH	ethanol
est	estimated
f	final
i	integral
in	in the feed
max	maximum

0	initial
O_2	oxygen
s	mass or standard
T	transition point
v	volumetric
vap	vaporization
w	water

Superscripts

i	i-species
max	maximum
o	oxygen
	dimensionless

10 References

1. Edwards, V. H. et al.: Biotech. Bioeng. *7*, 975 (1970)
2. Pirt, S. J.: Principles of Microbe and Cell Cultivation, p. 211, Oxford—London: Blackwell Scientific Publications 1975
3. Yoshida, F. et al.: Biotech. Bioeng. *15*, 257 (1973)
4. Burrows, S.: Baker's Yeast, in: The Yeast (eds. Rose, A. H., Harrison, J. S.), vol. 3, p. 350, New York: Academic Press 1970
5. Clark, D. J. et al.: Biochem. Biophys. Acta *92*, 85 (1964)
6. Hsu, E. J. et al.: J. Bacteriol. *98*, 172 (1969)
7. Yamane, K. et al.: J. Biochem. *67*, 9 (1970)
8. Hulme, M. A. et al.: Nature *226*, 469 (1970)
9. Kao, E. I.: Biotech. Bioeng. *18*, 1493 (1976)
10. Blanch, H. W. et al.: ibid. *18*, 748 (1976)
11. Yamanè, T.: Hakkokogaku *56*, 310 (1978)
12. Woehrer, W. et al.: Biotech. Bioeng. *23*, 567 (1981)
13. Wang, H. Y. et al.: ibid. *21*, 975 (1979)
14. White, J.: Yeast Technology, p. 31, New York: John Wiley Sons, Inc. 1954
15. Ohhashi, M.: Bull. Chem. Soc. Japan *61*, 1001 (1958)
16. Hospodka, J.: Biotech. Bioeng. *8*, 117 (1966)
17. Miskiewicz, T. et al.: ibid. *17*, 1829 (1975)
18. Aiba, S. et al.: ibid. *18*, 1001 (1976)
19. Wang, H. Y. et al.: ibid. *19*, 69 (1977)
20. Dairaku, K. et al.: ibid. *23*, 2069 (1981)
21. Dairaku, K. et al.: J. Ferment. Technol. *61*, 189 (1983)
22. Bach, H. P. et al.: Biotech. Bioeng. *20*, 799 (1978)
23. Nanba, A. et al.: J. Ferment. Technol. *59*, 383 (1981)
24. Harrison, D. E. F. et al.: Fermentation Technology Today (ed. Terui, G.), p. 491, Osaka: Society of Fermentation Technology, Japan 1972
25. Yamanè, T. et al.: J. Ferment. Technol. *54*, 229 (1976)
26. Srinivasen, V. R. et al.: Biotech. Bioeng. *19*, 153 (1977)
27. Minami, K. et al.: J. Ferment. Technol. *56*, 35 (1978)
28. Watteeuw, C. M. et al.: Biotech. Bioeng. *21*, 1221 (1979)
29. Reuss, M.: Eur. J. Appl. Microbiol. *1*, 295 (1975)
30. Nishio, N. et al.: J. Ferment. Technol. *55*, 151 (1977)
31. Yano, T. et al.: ibid. *56*, 416 (1978)
32. Kobayashi, T. et al.: Biotech. Bioeng. Symp. *9*, 73 (1979)
33. Mori, H. et al.: J. Chem. Eng. Japan *12*, 313 (1979)

34. Huang, S. Y. et al.: Biotech. Bioeng. *23*, 1491 (1981)
35. Yamanè, T. et al.: ibid. *23*, 2509 (1981)
36. Matsumura, M. et al.: J. Ferment. Technol. *60*, 565 (1982)
37. Moyer, J. A. et al.: J. Bact. *51*, 57 (1946)
38. Singh, K. et al.: ibid. *56*, 339 (1948)
39. Brown, W. E. et al.: Ind. Eng. Chem. *42*, 1769 (1950)
40. Brown, W. E. et al.: ibid. *42*, 1823 (1950)
41. Soltero, F. V. et al.: Appl. Microbiol. *1*, 52 (1953)
42. Anderson, R. F. et al.: Ind. Eng. Chem. *45*, 768 (1953)
43. Hosler, P. et al.: ibid. *45*, 872 (1953)
44. Davey, V. F. et al.: Appl. Microbiol. *1*, 208 (1953)
45. Soltero, F. V. et al.: ibid. *2*, 41 (1954)
46. Anderson, R. F. et al.: Agric. Food Chem. *4*, 556 (1956)
47. Noguchi, Y. et al.: J. Ferment. Technol. *38*, 511 (1960)
48. Hockenhull, D. J. D. et al.: Chem. Ind., p. 607 (1968)
49. Matelovà, V.: Appl. Microbiol. *23*, 669 (1972)
50. McCann, E. P. et al.: J. appl. Chem. Biotechnol. *22*, 1201 (1972)
51. Pan, C. H. et al.: Dev. Ind. Microbiol. *13*, 103 (1972)
52. Squires, R. W.: ibid. *13*, 128 (1972)
53. Calam, C. T. et al.: J. appl. Chem. Biotechnol. *23*, 225 (1973)
54. Fishman, V. M. et al.: Biotech. Bioeng. Symp. *4*, 647 (1974)
55. Hegewald, E. et al.: Biotech. Bioeng. *23*, 1563 (1981)
56. Mou, D.-G. et al.: ibid. *25*, 225, and 257 (1983)
57. Singh, A. et al.: Eur. J. Appl. Microbiol. *3*, 97 (1976)
58. Smith, C. G.: Appl. Microbiol. *4*, 232 (1956)
59. Calam, C. T. et al.: J. appl. Chem. Biotechnol. *21*, 181 (1971)
60. Martin, J. F. et al.: Dev. Ind. Microbiol. *15*, 324 (1974)
61. Matsumura, M. et al.: J. Ferment. Technol. *59*, 115 (1981)
62. Tanaka, K. et al.: J. Agric. Chem. Soc. Japan *34*, 593 (1960)
63. Su, Y.-C. et al.: Bull. Agr. Chem. Soc. Japan *24*, 525 (1960)
64. Tsunoda, T. et al.: J. Gen. Appl. Microbiol. *7*, 18 (1961)
65. Oki, T. et al.: Agr. Biol. Chem. *32*, 119 (1968)
66. Yamamoto, M. et al.: J. Ferment. Technol. *50*, 876 (1972)
67. Kishimoto, M. et al.: ibid. *59*, 43, and 125 (1981)
68. Yamada, H. et al.: Fermentation Technology Today (ed. Terui, G.), p. 445, The Society of Fermentation Technology Japan 1972
69. Ohno, H. et al.: Biotech. Bioeng. *18*, 847 (1976)
70. Aiba, S. et al.: Biotechnol. Letters *2*, 525 (1980)
71. Aiba, S. et al.: Appl. Environ. Microbiol. *43*, 289 (1982)
72. Fujio, Y. et al.: J. Ferment. Technol. *49*, 626 (1971)
73. Gray, P. P. et al.: Biotech. Bioeng. *15*, 1179 (1973)
74. Markkanen, P. et al.: J. appl. Chem. Biotechnol. *26*, 41 (1976)
75. Yamanè, T. et al.: J. Ferment. Technol. *55*, 233 (1977)
76. Suga, K. et al.: Proc. Bioconversion and Biochem. Eng. Symp. *2*, 371 (1980)
77. Allen, A. L. et al.: Biotech. Bioeng. *23*, 2641 (1981)
78. Gottvaldovà, M. et al.: Biotechnol. Letters *4*, 229 (1982)
79. Kojima, I. et al.: J. Ferment. Technol. *50*, 716 (1972)
80. Whitaker, A.: Process Biochem., May, p. 10 (1980)
81. Darken, M. A.: Appl. Microbiol. *7*, 301 (1959)
82. Masai, H. et al.: Bull. Agr. Chem. Soc. Japan *52*, R103 (1978)
83. Mori, H. et al.: J. Chem. Eng. Japan *14*, 65 (1981)
84. Yamauchi, H. et al.: J. Ferment. Technol. *61*, 275 (1983)
85. Endo, I. et al.: Proceeding of 2nd PACHEC, p. 583, Denver, USA 1977
86. Pirt, S. J.: J. appl. Chem. Biotechnol. *24*, 415 (1974)
87. Dunn, I. J. et al.: Biotech. Bioeng. *17*, 1805 (1975)
88. Yamanè, T. et al.: J. Ferment. Technol. *55*, 156 (1977)
89. Pirt, S. J.: Anal. N.Y. Acad. Sci. *326*, 119 (1979)

90. Dunn, I. J.: ibid. *326*, 127 (1979)
91. Yamanè, T. et al.: J. Ferment. Technol. *55*, 380 (1977)
92. Lim, H. C.: Biotech. Bioeng. *18*, 1635 (1976)
93. Jordan, R. C. et al.: J. Bact. *48*, 579 (1944)
94. Monod, J.: Ann. Rev. Microbiol. *3*, 371 (1949)
95. Kalogerakis, N. et al.: Biotech. Bioeng. *23*, 921 (1981)
96. Marr, A. G. et al.: Ann. N.Y. Acad. Sci. *102*, 536 (1979)
97. Lim, H. C. et al.: Biotech. Bioeng. *19*, 425 (1977)
98. Keller, R. et al.: J. appl. Chem. Biotechnol. *28*, 508 (1978)
99. Constantinides, A.: Ann. N.Y. Acad. Sci. *102*, 193 (1979)
100. Yamanè, T. et al.: J. Ferment. Technol. *55*, 587 (1977)
101. Dale, R. F. et al.: Appl. Microbiol. *1*, 68 (1953)
102. Martin, R. G.: Anal. Biochem. *8*, 43 (1964)
103. Pirt, S. J.: Biochem. J. *121*, 293 (1971)
104. Strohm, J. A. et al.: Ind. Eng. Chem. *53*, 760 (1961)
105. Yano, T. et al.: J. Ferment. Technol. *57*, 91 (1979)
106. Yano, T. et al.: ibid. *59*, 295 (1981)
107. Dietrich, K. R.: Ablaufverwertung und Abwasserreinigung, p. 280, A. Hüthig, Heidelberg 1959
108. Dairaku, K. et al.: Biotech. Bioeng. *21*, 1671 (1979)
109. Yano, T. et al.: J. Ferment. Technol. *56*, 421 (1978)
110. Yamanè, T. et al.: Biotech. Bioeng. *23*, 2493 (1981)
111. Kishimoto, M. et al.: Biotechnol. Letters *2*, 403 (1980)
112. Suzuki, T. et al.: Biotech. Bioeng. (to be submitted 1984)
113. Landwall, P. et al.: J. Gen. Microbiol. *103*, 345 (1977)
114. Keller, R. et al.: J. appl. Chem. Biotechnol. *28*, 784 (1978)
115. Shioya, S. et al.: J. Chem. Technol. Bioeng. *29*, 180 (1979)
116. Weigand, W. A.: Biotech. Bioeng. *23*, 249 (1981)
117. Mori, H. et al.: J. Ferment. Technol. *61*, 391 (1983)
118. Yamanè, T. et al.: Biotech. Bioeng. *21*, 111 (1979)
119. Mori, H. et al.: J. Ferment. Technol. *61*, 305 (1983)
120. Dairaku, K. et al.: ibid. *60*, 67 (1982)
121. Esener, A. A. et al.: Biotech. Bioeng. *23*, 1851 (1981)
122. Bajpai, R. K. et al.: ibid. *23*, 717 (1981)

Author Index Volumes 1—30

L.G.Nickell

Plant Growth Regulators

Agricultural Uses

1982. 29 figures. XII, 173 pages
ISBN 3-540-10973-0

Plant growth regulators can be defined as either natural or synthetic compounds that are applied directly to a target plant to alter its life processes or its structure to improve quality, increase yields, or facilitate harvesting.

The author of this book – Vice President of Research and Development at the Vesicol Chemical Corporation, Chicago, Chairman of the Plant Growth Regulator Working Group, and the Treasurer of the American Society of Plant Physiologists – discusses the effects of these regulators on plant functions such as rootinduction, control of flowering, control of sex, and control of maturation and aging. Emphasis is placed on the practical aspects of growth regulators rather than their mode of action. An extensive bibliography is provided which includes reviews by specialists in all areas of plant growth and development, both basic and applied.

Special consideration is given to the current commercial uses of plant growth regulators for a variety of purposes on a number of economically important crops as well as to the direct and indirect financial returns to the grower.

Tables are provided which show the chemical name, common name, trademark identification, code designation, producer companies, as well as plant growth regulatory activity and, where applicable, other biological activities of the regulators discussed in the text.

The book will be of interest to teachers, students and specialists in agronomy horticulture, plant physiology, crop science, nursery science, landscape architecture, pomology, seed technology, providing them with the only modern treatment of the uses of chemicals for the control of plant growth.
(1201 ref.)

Springer-Verlag
Berlin
Heidelberg
New York
Tokyo

GPSR Compliance

The European Union's (EU) General Product Safety Regulation (GPSR) is a set of rules that requires consumer products to be safe and our obligations to ensure this.

If you have any concerns about our products, you can contact us on ProductSafety@springernature.com

In case Publisher is established outside the EU, the EU authorized representative is:

Springer Nature Customer Service Center GmbH
Europaplatz 3
69115 Heidelberg, Germany

Batch number: 10620881

Printed by Printforce, the Netherlands